# Barron's Regents Exams and Answers

## Math B

D0696612

**LAWRENCE S. LEFF**
Former Assistant Principal
Mathematics Supervisor
Franklin D. Roosevelt High School
Brooklyn, New York

**DAVID BOCK**
Mathematics Instructor
Ithaca High School
Ithaca, New York

**THOMAS J. MARIANO**
Coordinator of Mathematics
Greece Central Schools
Rochester, New York

Barron's Educational Series, Inc.

*All inquiries should be addressed to:*
Barron's Educational Series, Inc.
250 Wireless Boulevard
Hauppauge, NY 11788
*http://www.barronseduc.com*

ISBN 0-7641-1728-9
ISSN 1528-1035

PRINTED IN THE UNITED STATES OF AMERICA
9 8 7 6 5 4 3 2 1

# Contents

1. Know What to Expect on Test Day
2. Avoid Last-Minute Studying
3. Be Well Rested and Come Prepared on Test Day
4. Know How to Use Your Calculator
5. Know When to Use Your Calculator
6. Have a Plan for Budgeting Your Time
7. Make Your Answers Easy to Read
8. Answer the Question That Is Asked
9. Take Advantage of Multiple-Choice Questions
10. Don't Omit Any Questions

# Graphing Calculator Skills     20

# Formulas     30

# Some Key Math B Facts and Skills     36

# Glossary of Terms     135

# Regents Examinations, Answers, and Self-Analysis Charts     151

# Preface

This book is designed to prepare you for the Mathematics B Regents examination while strengthening your understanding and mastery of the material on which this test is based.

In addition to providing actual Mathematics B Regents examinations, this book offers these special features:

- *Step-by-Step Solutions to All Regents Questions.* Careful study of the solutions and answer explanations will improve your mastery of the subject. Each explanation is designed to show you how to apply the facts and concepts you have learned in class. Since the explanation for each solution has been written with emphasis on the reasoning behind each step, its value goes well beyond the application to that particular question.

- *Unique System of Self-Analysis Charts.* Each set of solutions for a particular Regents examination ends with a Self-Analysis Chart and a classification of exam questions by topic. These charts will help you identify weaknesses and direct your study efforts where needed. The charts classify the questions on each Mathematics B Regents examination into an organized set of topic groups. This will also help you to locate other questions on the same topic from other Regents examinations.

- *General Test-Taking Tips.* Helpful tips and strategies are given that will help to raise your grade on the actual Regents exam that you take.

- *Graphing Calculator Skills.* The main features of the Texas Instruments TI-83 graphing calculator are reviewed. The procedures described, however, are similar for all graphing calculators.

- *Key Problems Organized by Major Topic.* Important facts and skills are reviewed through worked-out examples. The problems are organized under each of the major topics required by the New York State *Core Curriculum for Mathematics B*.

- *Glossary.* Definitions of important terms related to the Mathematics B Regents examination are conveniently organized in a glossary.

# Frequency of Topics— Math B

Questions in the Math B Regents exams fall into one of 30 topic categories. The "Some Key Math B Facts and Skills" that appears later in this book covers some of these concepts.

The Frequency Chart shows how many questions in recent exams have been in each category, to indicate which topics have been emphasized in earlier tests.

The Self-Analysis Charts that follow each Regents exam designate exactly which questions in that exam are in each category, so you can determine where your weakest areas are. You may also try questions in those areas again, for more practice.

The two charts—Frequency and Self-Analysis—should give you a very good idea of the topics you need to review, and the Practice Exercises in the "Facts and Skills" section provide more practice.

| | Number of Questions | | | | | | | |
|---|---|---|---|---|---|---|---|---|
| **Topic** | **Aug 2001** | **Jan 2002** | **June 2002** | **Aug 2002** | **June 2003** | **Aug 2003** | **June 2004** | **Aug 2004** |
| 1. Properties of Numbers | — | — | — | — | — | — | — | — |
| 2. Sequences | — | — | — | 1 | — | — | — | — |
| 3. Complex Numbers | — | 1 | 1 | 1 | 2 | 2 | 2 | 1 |
| 4. Inequalities, Absolute Value | 2 | 1 | 1 | 1 | 1 | 1 | — | 1 |
| 5. Algebraic Expressions, Fractions | 2 | 3 | 2 | 3 | 2 | 2 | 2 | 1 |
| 6. Exponents (zero, negative, rational, scientific notation) | — | 1 | — | 1 | 1 | 1 | 2 | 2 |
| 7. Radical Expressions | 1 | — | 2 | 1 | 1 | 1 | 2 | 2 |
| 8. Quadratic Equations (factors, formula, discriminant) | 2 | 4 | 2 | 5 | 1 | 2 | 1 | 3 |
| 9. Systems of Equations | 1 | 1 | 2 | — | 1 | 1 | — | — |
| 10. Functions (graphs, domain, range, roots) | 1 | 1 | 3 | 1 | 1 | 1 | 2 | 1 |
| 11. Inverse Functions, Composition | — | 2 | 2 | 2 | 1 | 2 | 1 | 1 |
| 12. Linear Functions | 3 | — | 1 | — | — | — | — | — |
| 13. Parabolas (max/min, axis of symmetry) | — | 1 | 1 | 1 | 1 | 1 | 2 | — |
| 14. Hyperbolas (including inverse variation) | 1 | 1 | — | 1 | 1 | 1 | 1 | 1 |
| 15. Ellipses | — | — | — | 1 | 1 | 1 | — | — |
| 16. Exponents and Logarithms | 2 | 3 | 2 | 3 | 4 | 2 | 2 | 1 |
| 17. Trig. (circular functions, unit circle, radians) | 4 | 4 | — | 1 | 3 | 2 | 1 | 4 |
| 18. Trig. Equations and Identities | 2 | — | 2 | — | 2 | 2 | 1 | 2 |
| 19. Solving Triangles (sin/cos laws, Pythag. thm., ambig. case) | 2 | 2 | 2 | 4 | 2 | 3 | 5 | 2 |
| 20. Coordinate Geom. (slope, distance, midpoint, circle) | — | — | — | — | 1 | — | — | 1 |
| 21. Transformations, Symmetry | 2 | 3 | 2 | 2 | 4 | 2 | 2 | 2 |

| | Number of Questions | | | | | | | |
|---|---|---|---|---|---|---|---|---|
| Topic | Aug 2001 | Jan 2002 | June 2002 | Aug 2002 | June 2003 | Aug 2003 | June 2004 | Aug 2004 |
| 22. Circle Geometry | 3 | 2 | 1 | 1 | — | 1 | 1 | 2 |
| 23. Congruence, Similarity, Proportions | — | — | 1 | — | 1 | 1 | — | 1 |
| 24. Probability (including Bernoulli events) | 1 | 1 | 1 | 1 | 2 | 1 | 1 | 2 |
| 25. Normal Curve | 1 | 1 | 1 | 2 | 1 | 1 | 2 | 1 |
| 26. Statistics (center, spread, summation notation) | 1 | — | 3 | 1 | 1 | — | 1 | 1 |
| 27. Correlation, Modeling, Prediction | 1 | 1 | 2 | 1 | — | 2 | 1 | 1 |
| 28. Algebraic Proofs, Word Problems | 2 | 1 | — | — | — | 1 | — | — |
| 29. Geometric Proofs | — | 1 | 1 | 1 | — | — | 2 | — |
| 30. Coordinate Geometry Proofs | 1 | — | — | — | 1 | — | — | 1 |

# How to Use This Book

As you work your way through this book, you will be following a carefully designed five-step study plan that will improve your understanding of the topics that the Mathematics B examination tests while also raising your Regents exam grade.

***Step 1. Know What to Expect on Test Day.*** Before the day of the test, you should be thoroughly familiar with the format, scoring, and special directions for the Mathematics B Regents exam. This will help you build confidence and prevent errors that may arise from not understanding the directions. The section titled "Getting Acquainted with Math B" gives you this important information.

***Step 2. Become Testwise.*** Reading the section titled "Ten Test-Taking Tips" will alert you to easy things you can do that will make you better prepared and more confident when you take the test.

***Step 3. Know How to Use Your Graphing Calculator.*** Become proficient in using your graphing calculator. Some questions on the Math B exam can be answered only by using a graphing calculator, while others become easier to answer when solved with a graphing calculator.

***Step 4. Review Specific Mathematics B Topics.*** You can gain skill and understanding in specific areas of Mathematics B by reviewing the key examples and their worked-out solutions which are organized by topic.

*Step 5. Take Practice Exams Under Exam Conditions.* The main part of this book contains actual Mathematics B Regents examinations with carefully worked-out solutions for all the questions. When you reach this part of the book, you should do these things:

• After you complete an exam, check the answer key for the entire test. Circle any omitted questions or questions you answered incorrectly and study the explained solutions for these questions.

• On the Self-Analysis Chart, find the topic under which each question is classified and enter the number of points you earned if you answered that question correctly.

• Figure out your percentage for each topic by dividing your earned points by the number of test points on that topic, carrying the division to two decimal places. If you are not satisfied with your percentage on any topic, reread the prepared solutions for the questions you missed. Then locate related questions in other Regents examinations by using the Self-Analysis Chart to see which questions are listed for the troublesome topic. Attempting to solve these questions and then studying their solutions will provide you with additional help. You may also find it helpful to review the appropriate sections in "Some Key Math B Facts and Skills." If you need more detailed explanations and additional practice problems with answers, you may want to get Barron's companion book, *Let's Review: Math B.*

# Tips for Practicing Effectively and Efficiently

- Do not spend too much time on any one question. If you cannot come up with a method to use, or if you cannot complete the solution, put a slash through the number of the question. When you have completed as many questions as you can, return to these questions and try them again.

- After trying the unanswered questions again, compare each of your solutions with the solutions that are provided. Read the explanations given even if you have answered the question correctly. Each solution has been carefully designed to provide additional insight into the topic that may be valuable when answering a more difficult question on the same topic on the next test you take.

- In the weeks before the day of the actual test, you should plan to spend at least $\frac{1}{2}$ hour preparing each night. It is better to spread out your preparation time this way than to cram by preparing for, say, 3 hours, in one evening. As the day of the Regents exam approaches, take at least one Regents exam under test conditions.

# Getting Acquainted with Math B

## WHAT IS MATH B?

The term "Math B" refers to a new Regents examination in mathematics. The Mathematics B Regents examination was offered for the first time in June 2000 and will eventually replace the Regents examination in Course III. If you entered high school in September 2000 or later, you must pass either the Mathematics B or the Course III Regents examination in order to qualify for an advanced Regents diploma.

## WHEN DO I TAKE THE MATHEMATICS B REGENTS EXAM?

The Mathematics B Regents exam is administered in January, June, and August of every school year. Most students take this exam after completing approximately 3 years of high-school-level mathematics courses. Your school and the mathematics program in which you are enrolled determine when you should take the Mathematics B Regents exam.

## HOW IS THE MATHEMATICS B REGENTS EXAM SET UP?

The Mathematics B Regents exam is a 3-hour exam that consists of four parts with a total of 34 questions. The accompanying table shows the point breakdown of the test.

| Part | Number of Questions | Point Value | Total Points |
|------|--------------------|-------------|--------------|
| I | 20 multiple choice | 2 each | $20 \times 2 = 40$ |
| II | 6 | 2 each | $6 \times 2 = 12$ |
| III | 6 | 4 each | $6 \times 4 = 24$ |
| IV | 2 | 6 each | $2 \times 6 = 12$ |
| | Test = 34 questions | | Test = 88 points |

- Part I consists of 20 standard multiple-choice questions with four possible answers labeled (1), (2), (3), and (4).
- The questions in Parts II, III, and IV require that you show or explain how you arrived at your answer by indicating the necessary steps involved, including appropriate formula substitutions, diagrams, graphs, charts, and so forth. If you use a guess-and-check strategy to arrive at an answer for a problem, you must include at least one guess that does *not* work before you show the correct guess.
- All the questions in each of the four parts of the test must be answered.

## WHERE DO I SHOW MY ANSWERS AND WORK?

Scrap paper is not provided or permitted for any part of the exam.

- After you figure out the answer to each multiple-choice question in Part I, you must write the numeral that precedes the correct choice in the space provided on the separate tear-off answer sheet for Part I found at the back of the question booklet.
- The answers and the work for the questions in Parts II, III, and IV must be written directly in the question booklet in the space provided underneath the questions. All work should be written in pen except for graphs which should be drawn in pencil.
- If you need graph paper, it will be provided in the question booklet.

# HOW IS YOUR FINAL REGENTS SCORE DETERMINED?

Solutions to questions in Parts II, III, and IV that are not completely correct may receive partial credit according to a special scoring guide provided by the New York State Education Department.

The maximum total raw score for the Mathematics B Regents exam is 88 points. After the raw scores for the four parts of the test are added together, a conversion table provided by the New York State Education Department is used to convert your raw score to a final test score that falls within the usual 0-to-100 scale.

# WHAT TYPE OF CALCULATOR DO I NEED?

Graphing calculators are *required* for the Mathematics B Regents examination. A graphing calculator allows you to quickly create a table of values or visualize a problem that can be represented by an equation or an inequality. Some problems can be solved more easily graphically with a calculator than algebraically with pen and paper. Furthermore, a graphical solution with a calculator may help to confirm an answer obtained using standard algebraic methods.

Here are some of the skills you will need to master in working with your graphing calculator:

- Find powers and roots of numbers.
- Evaluate permutations $(_nP_r)$, combinations $(_nC_r)$, logarithms, and trigonometric functions.
- Graph linear, quadratic, exponential, logarithmic, and trigonometric functions in an appropriate viewing window so that their key features can be seen.
- Solve equations and systems of equations graphically by finding $x$-intercepts and points of intersection.
- Calculate the mean and standard deviation of a set of data values.
- Determine the line or curve that best fits paired data and then use the regression equation for interpolation and extrapolation.

- Create a scatterplot and determine the coefficient of linear correlation between paired data.

When you use your graphing calculator to draw a graph as part of an answer, you are expected to show each of the following:

- An accurate sketch of the viewing window.
- Scales indicated on the coordinate axes.
- Clearly labeled $x$- and $y$-intercepts and coordinates of points of intersection, if needed for the solution.

When answering statistics-related questions with a graphing calculator, you will need to show the key elements of your solution.

- For *standard deviation questions*, show the number of scores, the mean, and the standard deviation.
- For *regression questions*, write the regression equation and, if needed, show any required substitutions for $x$ in order to obtain the predicted values for $y$.
- For *coefficient of linear correlation questions*, show the regression equation and indicate how the coefficient of correlation was used in your solution.

You can find additional information about how specific types of problems can be solved using a graphing calculator in Barron's companion review book, *Let's Review: Math B*.

## WHAT IS COLLECTED AT THE END OF THE EXAMINATION?

At the end of the examination, you need to return:

- Any tool provided to you by your school such as a compass or a graphing calculator.
- The question booklet. Check that you have printed your name and the name of your school in the appropriate boxes near the top of the first page of the question booklet.

- The Part I answer sheet. You must sign the statement at the bottom of the Part I answer sheet indicating that you did not receive unlawful assistance in answering any of the questions. If you fail to sign this declaration, your answer paper will not be accepted.

## WHAT IS THE *CORE CURRICULUM*?

The *Core Curriculum* is the official publication of the New York State Education Department that lists the topics that can be tested by the Mathematics B Regents examination. This examination can test you on a wide range of topics that includes:

- Direct, indirect, and coordinate geometric proofs.
- Area and volume.
- Circle relationships and angle measurement.
- Properties and operations with real and complex numbers; algebraic operations with fractions and radicals; factoring.
- Solving quadratic equations including those with irrational and nonreal roots; solving systems of equations.
- Linear, quadratic, logarithmic, exponential, and trigonometric functions and their graphs.
- Transformations and functions.
- Normal curve; fitting a line or curve to data using least squares regression; scatterplots; coefficient of linear correlation.
- Counting methods and probability in two-outcome experiments.
- Trigonometric equations and laws.

If you have Internet access, you can view the *Core Curriculum* on the New York State Education Department's web site at *http://www.emsc.nysed.gov/ciai/pub.html#cat4*.

# Ten Test-Taking Tips

Here are ten practical tips that can help you raise your grade on the Mathematics B Regents examination.

---

**TIP 1**

**Know What to Expect on Test Day**

---

## SUGGESTIONS
- Become familiar with the format and directions for the Mathematics B Regents exam.
- Know where you should write your answers for the different parts of the exam.
- Ask your teacher to show you an actual test booklet for a previously given Mathematics B Regents exam.

---

## TIP 2

**Avoid Last-Minute Studying**

---

## SUGGESTIONS

- Start your Regents exam preparation early by making a regular practice of: taking detailed notes in class and then reviewing them when you get home; completing all written homework assignments in a neat, organized way; writing down any questions you may have about your homework so that you can ask your teacher about them; saving your classroom tests so that you can use them as an additional source of questions when preparing for the Regents exam.

- Build skill and confidence by completing all the exams in this book and studying the accompanying solutions before the day of the Regents exam. Because each exam takes up to 3 hours to complete, you should begin this process no later than several weeks before the exam is scheduled to be given.

- Get a review book early in your exam preparation so that if you need additional help or explanations, it will be at your fingertips. The recommended review book is Barron's *Let's Review: Math B*. This easy-to-follow book has been designed for fast, effective learning.

- As the day of the actual Regents exam gets closer, take the exams in this book under timed examination conditions. Then compare each of your answers with the explained answers contained in the book.

- Use the Self-Analysis Chart at the end of each exam to help pinpoint any weaknesses.

- If you do not feel confident in a particular area, study the corresponding topic in Barron's *Let's Review: Math B*.

- As you work your way through the exams in this book, make a list of any formulas or rules that you need to know. Learn this material well before the day of the exam.

---

## TIP 3

## Be Well Rested and Come Prepared on Test Day

---

## SUGGESTIONS

- On the night before exam day, lay out all the things you must bring to the exam room.
- Prepare a checklist to make sure you bring these items:

  ☐ Regents admission card with the room number of the exam.
  ☐ Two ink pens.
  ☐ Two sharpened pencils with erasers.
  ☐ A ruler.
  ☐ A compass.
  ☐ A graphing calculator that you know how to use. If you do not own one, your school is required to make one available to you during the examination.
  ☐ A watch.

- If your calculator uses batteries, put fresh ones in it the night before the exam.
- Eat wisely and go to bed early so you will be alert and well rested when you take the exam.
- Be certain you know when your exam begins. Set an alarm clock to give yourself plenty of time to arrive at school before the exam starts.
- Tell your parents what time you will need to leave the house in order to get to school on time.
- Arrive at the exam room confident by being on time and being well prepared.

---

## TIP 4

### Know How to Use Your Calculator

## SUGGESTIONS

- The graphing calculator you take to the exam room should be the same calculator you used when you completed the practice Regents exams at home.
- If you are required to use a graphing calculator provided by your school, make sure you practice with it before the day of the exam because not all graphing calculators work the same way.

## TIP 5

### Know When to Use Your Calculator

## SUGGESTIONS

- Don't expect to have to use your calculator to help answer each question.
- If you need to create a table of values in order to graph a parabola, use the table-building feature of your calculator.
- Expect to have to use your calculator when solving numerical problems involving trigonometric, exponential, and logarithmic functions.
- Know how to use the special statistical features of your calculator to obtain standard deviation values and regression equations.
- Get into the habit of using your calculator to evaluate factorials, permutations, and combinations because these calculations are prone to error when performed manually using the appropriate formulas.

---

## TIP 6

### Have a Plan for Budgeting Your Time

---

## SUGGESTIONS

- In the first **60 minutes** of the 3-hour exam, complete the 20 multiple-choice questions in Part I. In answering troublesome multiple-choice questions, first rule out any choices that are impossible. If the answer choices are numbers, you may be able to identify the correct answer by plugging these numbers back into the original question to see if they work. If the answers choices are letters, you can substitute easy numbers for the letters both in the test question and in each of the answer choices. You can then try to match the number produced by the question to the answer choice that evaluates to the same number. This is explained more fully in Tip 9.

- In the next **70 minutes** of the exam, complete the six Part II questions and the six Part III questions. In order to maximize your credit for each question, clearly write down the steps you followed to arrive at each answer. Include any equations, formula substitutions, diagrams, tables, graphs, and so forth.

- In the last **50 minutes** of the exam:

  1. Complete the two Part IV questions. Again, be sure to show how you arrived at each answer.
  2. During the last 10 minutes, review your entire test paper for neatness and accuracy. Check that all answers (except for graphs) are written in ink. Make sure you have answered all the questions in each part of the exam and that all your Part I answers have been recorded accurately on the separate Part I answer sheet.
  3. Before you submit your test materials to the proctor, check that you have written your name in the reserved spaces on the front page of the question booklet and on the Part I answer sheet. Don't forget to sign the declaration that appears at the bottom of the Part I answer sheet.

## TIP 7

### Make Your Answers Easy to Read

## SUGGESTIONS

- Make sure your answers are clear, neat, and written in ink (except for graphs). When solving problems algebraically, define what the variables stand for, as in "Let $x = \ldots$"
- Draw graphs using a pencil so that you can erase neatly, if necessary. Use a ruler to draw straight lines and axes.
- Don't forget to label the coordinate axes. Put a "$y$" on top of the vertical axis and a "$-y$" on the bottom. Write "$x$" to the right of the horizontal axis and "$-x$" to the left. Next to each graph, write its equation.
- When answering a question in Part II, III, or IV, record your reasoning as well as your final answers. Provide enough details so that it will be clear to someone who doesn't know how you think, why and how you moved from one step of the solution to the next. If it is difficult for the teacher who is grading your paper to figure out what you have written, the teacher may simply decide to mark your work as incorrect and give you little, if any, partial credit.
- Draw a box around your final answer to a Part II, III, or IV question.

## TIP 8

### Answer the Question That Is Asked

## SUGGESTIONS

- Make sure your answer is in the form required by the question. For example, if a question asks for an approximation, round off your answer to the required decimal position. If a question asks that you write an answer in lowest terms, make sure that the numerator and denominator of a fractional answer do not have any common factors other than 1 or −1. For example, instead of leaving the answer as $\frac{10}{12}$, write $\frac{5}{6}$; instead of leaving the answer as $\frac{x^2-1}{x+1}$, write $x - 1$ since

$$\frac{x^2-1}{x+1} = \frac{(x-1)\overset{1}{\cancel{(x+1)}}}{\cancel{x+1}} = x - 1$$

If the question calls for the answer in simplest radical form, make sure you simplify a square root radical so that the radicand does not contain any perfect square factors greater than 1. For example, instead of leaving the answer as $\sqrt{18}$, write it as $3\sqrt{2}$.

- If the question asks for the $x$-coordinate (or $y$-coordinate) of a point, do not give both the $x$- *and* the $y$-coordinates.
- After solving a word problem, check back in the original question to make sure your *final* answer works and is the particular quantity the question asks you to find.
- If units of measurement are given, as in area problems, check that your answer includes the correct units of measurement.
- If the question requires a positive root of a quadratic equation, as in geometric problems in which the variable represents a dimension, make sure you reject the negative root.

---

### TIP 9

**Take Advantage of Multiple-Choice Questions**

---

## SUGGESTIONS
- When the answer choices for a multiple-choice question contain only numbers, try plugging each answer choice into the original question until you find the one that works.

• If the answer choices for a multiple-choice question contain only letters, replace the letters with easy numbers both in the question and in the answer choices. Then work out the problem using these numbers.

**EXAMPLE**

What is the solution set of the inequality $x^2 + 3x - 10 > 8$?

(1) $\{x|-6 < x < 3\}$                    (3) $\{x|-3 < x < 6\}$

(2) $\{x|x < -6 \text{ or } x > 3\}$              (4) $\{x|x < -3 \text{ or } x > 6\}$

**Solution 1:** Use algebra.

• First find the roots of the related quadratic equation. If $x^2 + 3x - 10 = 8$, then $x^2 + 3x - 18 = 0$; so $(x + 6)(x - 3) = 0$. Hence, $x = -6$ or $x = 3$.

• In general, the solution of a quadratic inequality $x^2 + bx + c > 0$ has the form $\{x|x < r \text{ or } x > R\}$, where $r$ and $R$ are the roots of the related quadratic equation with $r < R$.

• Since $r = -6$ and $R = 3$, the solution set is $\{x|x < -6 \text{ or } x > 3\}$; so the correct choice is **(2)**.

**Solution 2:** Substitute numbers.

• Since 0 is a member of the solution set given in choice (1), check whether it satisfies the original inequality:

$$0^2 + 3 \cdot 0 - 10 > 8$$
$$-10 > 8 \qquad \text{False}$$

Eliminate choice (1). For the same reason, eliminate choice (3).

• Since −7 is a member of the solution set given in choice (3), check whether it satisfies the original inequality:

$$(-7)^2 + 3(-7) - 10 > 8$$
$$49 \quad - 21 \quad - 10 > 8$$
$$9 > 8 \qquad \text{True}$$

• Hence, the correct choice is **(2)**.

**EXAMPLE**

What is the value of $\sin\left(\arccos\dfrac{1}{x}\right)$?

(1) $\dfrac{\sqrt{1-x^2}}{x}$    (2) $\dfrac{\sqrt{1+x^2}}{x}$    (3) $\dfrac{\sqrt{x^2-1}}{x}$    (4) $\dfrac{x}{\sqrt{x^2+1}}$

**Solution 1:** Analyze using trigonometric relationships.

The notation $\sin\!\left(\arccos\dfrac{1}{x}\right)$ is read as "the sine of the angle whose cosine is $\dfrac{1}{x}$." Let $\theta$ represent the Quadrant I angle whose cosine is $\dfrac{1}{x}$.

- Draw a reference triangle in which $\cos\theta = \dfrac{1}{x}$, as shown in the accompanying diagram.

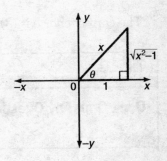

- Use the Pythagorean theorem to represent the vertical side, $y$, of the triangle in terms of $x$:

$$1^2 + y^2 = x^2$$
$$y^2 = x^2 - 1$$
$$y = \sqrt{x^2 - 1}$$

- Thus:

$$\sin\!\left(\arccos\dfrac{1}{x}\right) = \sin\theta = \dfrac{\text{side opposite } \theta}{\text{hypotenuse}}$$
$$= \dfrac{\sqrt{x^2-1}}{x}$$

The correct choice is **(3)**.

**Solution 2:** Pick an easy number for $x$.

If $x = 2$, then $\sin\!\left(\arccos\dfrac{1}{2}\right) = \sin 60° = \dfrac{\sqrt{3}}{2}$. Next, substitute 2 for $x$ in each of the answer choices until you find the one that evaluates to $\dfrac{\sqrt{3}}{2}$:

- Choice (1): $\dfrac{\sqrt{1-x^2}}{x} = \dfrac{\sqrt{1-2^2}}{2} = \dfrac{\sqrt{-3}}{2}$. Reject.

- Choice (2): $\dfrac{\sqrt{1+x^2}}{x} = \dfrac{\sqrt{1+2^2}}{2} = \dfrac{\sqrt{5}}{2}$. Reject.

- Choice (3): $\dfrac{\sqrt{x^2-1}}{x} = \dfrac{\sqrt{2^2-1}}{2} = \dfrac{\sqrt{3}}{2}$. Correct—so pick this choice.

- Choice (4): $\dfrac{x}{\sqrt{x^2+1}} = \dfrac{2}{\sqrt{2^2+1}} = \dfrac{2}{\sqrt{5}}$. Reject.

The correct choice is **(3)**.

---

## TIP 10

### Don't Omit Any Questions

---

## SUGGESTIONS

- Keep in mind that in each of the four parts of the test you must answer all the questions.
- If you get stuck on a multiple-choice question in Part I, try one of the problem-solving strategies discussed under Tip 9. If you still can't figure out the answer, try to eliminate any impossible answers. Then guess one of the remaining answer choices.
- Try to maximize your partial credit for a Part II, III, or IV question by writing down any formula, diagram, or mathematics facts you think might apply. If appropriate, organize and analyze the given information by making a table or a diagram. You can then try to arrive at the correct answer by guessing, checking your guess, and then revising your guess, as needed.
- If you get stuck on a question from Part II, III, or IV, use the general problem-solving approach summarized in the box on page 19.

# A General Problem-Solving Approach

1. ***Read*** each test question through the *first time* to get a *general idea* of the type of mathematics knowledge the question requires. For example, one question may ask you to solve an algebraic equation, while another question may ask you to apply some geometric principle. Then ***read*** the problem through a *second time* to pick out specific facts. Identify what is *given* and what you need to *find*.

2. ***Decide how you will solve the problem.*** You may need to use one of the special problem-solving strategies discussed in this section. If you decide to solve a word problem by using algebraic methods, you may first need to translate the conditions of the problem into an equation or an inequality.

3. ***Carry out your plan.*** Show the steps you followed in arriving at your answer.

4. ***Verify that your answer is correct*** by making sure it works in the original question.

# Summary of Test-Taking Tips

1. Know what to expect on test day.
2. Avoid last-minute studying.
3. Be well rested and come prepared on test day.
4. Know how to use your calculator.
5. Know when to use your calculator.
6. Have a plan for budgeting your time.
7. Make your answers easy to read.
8. Answer the question that is asked.
9. Take advantage of multiple-choice questions.
10. Don't omit any questions.

# Graphing Calculator Skills

Because of its popularity and availability, the Texas Instruments TI-83 graphing calculator will be used as the "reference" calculator. if you are using a different graphing calculator, you may need to make minor adjustments in the instructions so that they will work for your type of calculator. If necessary, consult the manual that came with your calculator.

## 1. PERFORMING CALCULATIONS

Routine arithmetic calculations are performed in the "home screen." Enter the home screen by pressing $\boxed{\text{2nd}}$ $\boxed{\text{MODE}}$. After you enter an arithmetic expression, press $\boxed{\text{ENTER}}$ to see the answer. The result is displayed at the end of the next line, as shown in accompanying screen shot. Notice that the calculator key for exponentiation is $\boxed{\wedge}$.

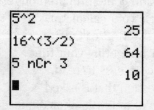

Unlike a scientific calculator, the TI-83 does not have special keys for evaluating combinations and permutations. Instead, you have to access the calculator's math library of special functions. To evaluate $_5C_3$ (or $_5P_3$) using a TI-83 calculator:

- Enter 5.
- Press MATH ▷ ▷ ▷ 3 to select $_nC_r$ (or use option 2 to select $_nP_r$) from the MATH PRB menu.
- Enter 3 and press ENTER. The result is 10.

## 2. GRAPHING FUNCTIONS AND SETTING WINDOWS

Before you can graph the function $y = x^2 - 6x + 8$, you must enter it in the Y = editor. Variable $x$ is entered by pressing $x, T, \theta, n$. Notice that it is possible to enter more than one equation.

Set an appropriate viewing window by pressing ZOOM and then selecting option 4 for a decimal window or option 6 for a standard window. In a basic decimal window, the axes are scaled so that $-4.7 \leq x \leq 4.7$ and $-3.1 \leq y \leq 3.1$. This allows the screen cursor to move in "friendly" steps of 0.1. In a standard window, the coordinate axes are scaled so that there are 10 tick marks on either side of the origin on both axes.

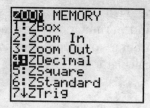 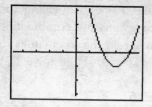

If you find that the graph does not fit in a basic decimal window, you can multiply the screen dimensions by a whole number by pressing  $\boxed{\text{WINDOW}}$  and then multiplying the values of Xmin and Xmax by the same whole number. The next set of screen shots show the graph of $Y_1 = x^2 - 6x + 8$ in a friendly window in which the basic decimal values of Xmin and Xman have each been multiplied by 2:

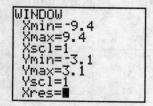 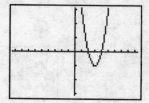

## 3. CREATING TABLES

If you want to graph $Y_1 = x^2 - 6x + 8$ using graph paper, you will need a table of values. After the equation of the parabola has been entered in the Y = editor press $\boxed{\text{2nd}}$ $\boxed{\text{GRAPH}}$. A table of values will be displayed in which the value of $x$ increases in steps of 1.

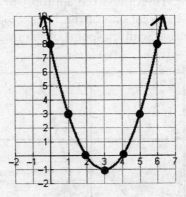

To see table values not in the current window, press the up or down cursor key. If you want $x$ to increase in increments other than 1, press $\boxed{\text{2nd}}$ $\boxed{\text{WINDOW}}$. To increase $x$ in steps of 0.5, make $\Delta$Tb1 = 0.5.

## 4. FINDING INTERSECTION POINTS

When two graphs intersect, you can find the coordinates of their point(s) of intersection using the TRACE or INTERSECT feature of your graphing calculator. To solve the system of equations $y = 3x - 5$ and $y = x - 1$ graphically, enter both equations in the Y = editor. Then graph these equations in a decimal window.

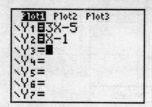 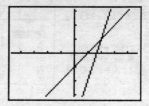

- Press TRACE. Then press the appropriate cursor key so that the cursor moves along one of the lines. Stop when the cursor appears to coincide with the point of intersection. The readout will be $x = 2, y = 1$.

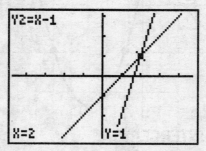

To verify that the point (2,1) is the point of intersection, toggle between the two curves by pressing the up or down cursor key. The equation in the top left corner of the screen indicates the graph on which the cursor is currently located. As the cursor position changes from one graph to the other, the equation in the top left corner of the screen changes, but the coordinates at the bottom of the screen remain set at $x = 2$ and $y = 1$. This confirms that the point (2,1) is common to both lines and is therefore the point of intersection.

- You can also obtain the point of intersection by pressing 2nd TRACE and then selecting option **5:intersect** from the CALCULATE menu. Move the cursor to a point on the first curve that is close to the point of intersection and press ENTER. Then move the cursor on the second curve to a point that is close to the point of intersection and press ENTER two times. The coordinates of the point of intersection will appear at the bottom of the screen.

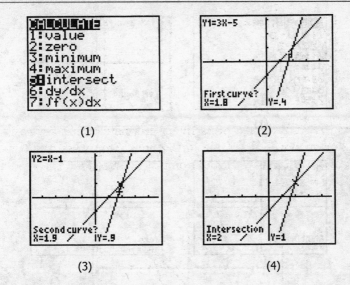

(1)  (2)

(3)  (4)

## 5. FINDING x-INTERCEPTS

To find an x-intercept or zero of the graph of $y = x^2 - 4x + 2$, select **2:zero** from the CALCULATE menu. Move the cursor to a point on the graph that is slightly to the left of the first x-intercept. Press ENTER. Then move the cursor to a point on the graph that is slightly to the right of the same x-intercept. Press ENTER two times to display the coordinates of that x-intercept.

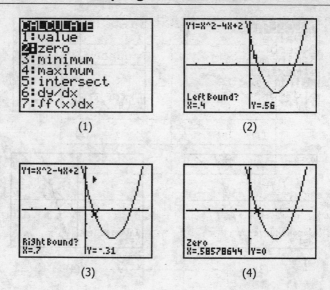

(1)            (2)

(3)            (4)

Hence, one of the irrational roots of the equation $x^2 - 4x + 2 = 0$ is, to the *nearest hundredth*, 0.59.

## 6. GRAPHING TRIGONOMETRIC FUNCTIONS

To graph a trigonometric function such as $y = 3\sin 2x$, enter the function in the Y = editor and then press $\boxed{\text{ZOOM}}$ $\boxed{7}$. If the angular mode is set to radians, the graph will be displayed in the interval $-2\pi < x < 2\pi$ using the following preset values:

$$\text{Xmin} = -\left(\frac{47}{24}\right)\pi, \text{Xmax} = \left(\frac{47}{24}\right)\pi, \text{Xscl} = \frac{\pi}{2}, \text{ and Yscl} = 1$$

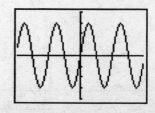

# 7. CALCULATING STANDARD DEVIATION

| Test Score, $x_i$ | Frequency, $f_i$ |
|:---:|:---:|
| 80 | 5 |
| 85 | 7 |
| 90 | 9 |
| 95 | 4 |

To calculate the standard deviation of the data in the accompanying table, press STAT ENTER. Enter the set of $x$ data values in list L1 and the corresponding frequencies in list L2.

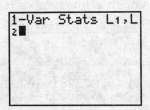

Press STAT ▷ ENTER. On the display line that begins "1-Var Stats," press 2nd 1 , 2nd 1 .

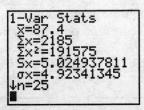

Press ENTER to get the standard deviation $\sigma_x \approx 4.92$.

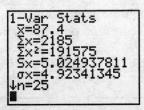

## 8. FITTING A REGRESSION LINE TO DATA

The line or curve (exponential, power, or logarithmic) that best fits a paired set of data values can be obtained using the regression feature of a graphing calculator.

| Minutes Studied ($x$) | 15 | 40 | 45 | 60 | 70 | 75 | 90 |
|---|---|---|---|---|---|---|---|
| Test Grade ($y$) | 50 | 67 | 75 | 75 | 73 | 89 | 93 |

If the $x$-values in the accompanying table have already been stored in list L1 and the $y$-values stored in list L2, the regression line or line of best fit can be calculated. Press $\boxed{\text{STAT}}$ $\boxed{\triangleright}$ $\boxed{4}$ to choose the **LinReg**(ax+b) option. Then press $\boxed{\text{ENTER}}$ to show a summary of the regression statistics when $a$ is the slope of the regression line, $b$ is the $y$-intercept, and $r$ is the coefficient of linear correlation. An equation of the line of best fit with the regression coefficients rounded off to the *nearest hundredth* is $y = 0.53x + 44.70$. Since the coefficient of linear correlation, $r$, is close to 1, the line is a good fit to the data.

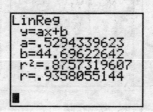

```
LinReg
 y=ax+b
 a=.5294339623
 b=44.69622642
 r²=.8757319607
 r=.9358055144
■
```

- You can store the regression equation as $Y_1$ if immediately after selecting the linear regression option, you press:

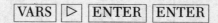

$\boxed{\text{VARS}}$ $\boxed{\triangleright}$ $\boxed{\text{ENTER}}$ $\boxed{\text{ENTER}}$

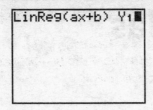

Press $\boxed{Y=}$ to see the regression equation.

• To predict the value of $y$ when $x = 80$ minutes, substitute 80 for $x$ in the regression equation:

$$y = 0.53x + 44.70 = 0.53 \times 80 + 44.7 \approx 87$$

If the regression equation has been stored as $Y_1$, you can also obtain the predicted $y$-value by using the table feature of your calculator:

| X | Y1 | |
|---|---|---|
| 75 | 84.404 | |
| 76 | 84.933 | |
| 77 | 85.463 | |
| 78 | 85.992 | |
| 79 | 86.522 | |
| 80 | 87.051 | |
| 81 | 87.58 | |

X=80

# Formulas

The following formulas are provided in the Mathematics B question booklet.

**Area of Triangle**

$$K = \frac{1}{2} ab \sin C$$

**Functions of the Sum of Two Angles**

$$\sin(A + B) = \sin A \cos B + \cos A \sin B$$
$$\cos(A + B) = \cos A \cos B - \sin A \sin B$$

**Functions of the Difference of Two Angles**

$$\sin(A - B) = \sin A \cos B - \cos A \sin B$$
$$\cos(A - B) = \cos A \cos B + \sin A \sin B$$

**Law of Sines**

$$\frac{a}{\sin A} = \frac{b}{\sin B} = \frac{c}{\sin C}$$

**Law of Cosines**

$$a^2 = b^2 + c^2 - 2bc \cos A$$

**Functions of the Double Angle**

$$\sin 2A = 2 \sin A \cos A \qquad \cos 2A = \cos^2 A - \sin^2 A$$
$$\cos 2A = 2 \cos^2 A - 1 \qquad \cos 2A = 1 - 2 \sin^2 A$$

**Functions of the Half-Angle**

$$\sin \frac{1}{2} A = \pm\sqrt{\frac{1 - \cos A}{2}} \qquad \cos \frac{1}{2} A = \pm\sqrt{\frac{1 + \cos A}{2}}$$

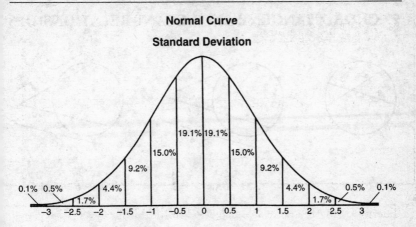

**Normal Curve**

**Standard Deviation**

# SOME ADDITIONAL FORMULAS YOU SHOULD KNOW

## 1. ANGLE MEASUREMENT IN CIRCLES

| Location of Vertex of Angle | Measure of Angle Equals |
| --- | --- |
| Center of circle | The measure of the intercepted arc |
| On circle | One-half the measure of the intercepted arc |
| Inside circle | One-half the sum of the measures of the intercepted arcs |
| Outside circle | One-half the difference of the measures of the intercepted arcs |

## 2. CHORD, TANGENT, AND SECANT RELATIONSHIPS

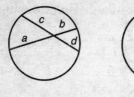 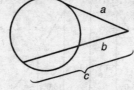 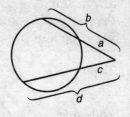

$a \times b = c \times d$ $\qquad$ $a^2 = b \times c$ $\qquad$ $a \times b = c \times d$

## 3. CIRCUMFERENCE, AREA, AND VOLUME

| Quantity | Formula |
|---|---|
| Circumference | $2\pi \times$ Radius |
| Length of arc intercepted by a central angle of $n°$ or $\theta$ radians | $\dfrac{n}{360} \times$ Circumference *or* radius $\times \theta$ |
| Area of circle | $\pi \times (\text{Radius})^2$ |
| Area of sector formed by a central angle of $n°$ | $\dfrac{n}{360} \times$ Area |
| Volume of a rectangular solid | Length $\times$ width $\times$ height |

## 4. COORDINATE FORMULAS

| Property of Segment Joining $A(x_A, y_A)$ and $B(x_B, y_B)$ | Formula or Relationship |
|---|---|
| Slope | $\dfrac{y_B - y_A}{x_B - x_A}$ |
| Midpoint | $\left(\dfrac{x_A + x_B}{2}, \dfrac{y_A + y_B}{2}\right)$ |
| Length | $\sqrt{(x_B + x_A)^2 + (y_B - y_A)^2}$ |
| Equation of line: slope-intercept | $y = mx + b$, where $m$ is the slope of the line and $b$ is the $y$-intercept |

## 5. RIGHT-TRIANGLE PROPORTIONS

| Altitude Drawn to Hypotenuse | Proportions Formed |
|---|---|
| | $\dfrac{x}{h} = \dfrac{h}{y}$ <br><br> $\dfrac{c}{a} = \dfrac{a}{y}$ <br><br> $\dfrac{c}{b} = \dfrac{b}{x}$ |

# 6. QUADRATIC EQUATIONS

| Feature of $ax^2 + bx + c = 0$ $(a \neq 0)$ | Formula |
| --- | --- |
| Two roots | $x = \dfrac{-b \pm \sqrt{b^2 - 4ac}}{2a}$ |
| Sum of roots | $-\dfrac{b}{a}$ |
| Product of roots | $\dfrac{c}{a}$ |

# 7. LOGARITHM LAWS

| Name | Law |
| --- | --- |
| Product | $\log(xy) = \log x + \log y$ |
| Quotient | $\log\left(\dfrac{x}{y}\right) = \log x - \log y$ |
| Power | $\log(x^y) = y \log x$ |
| Change of base | $y = \log_b x = \dfrac{\log x}{\log b}$ |

# 8. TANGENT IDENTITIES

| Angle Relationship | Identity |
|---|---|
| Sum of two angles | $\tan(A + B) = \dfrac{\tan A + \tan B}{1 - \tan A \tan B}$ |
| Difference of two angles | $\tan(A - B) = \dfrac{\tan A - \tan B}{1 + \tan A \tan B}$ |
| Double angle | $\tan 2A = \dfrac{2 \tan A}{1 - \tan^2 A}$ |
| Half-angle | $\tan \dfrac{1}{2} A = \pm \sqrt{\dfrac{1 - \cos A}{1 + \cos A}}$ |

# 9. COMBINATIONS AND COUNTING

| | Formula |
|---|---|
| The number of ways in which a subcommittee of $r$ members can be selected from $n$ members where $r \leq n$ | $_nC_r = \dfrac{n!}{(n - r)! \cdot r!}$ |
| Probability of $r$ successes in $n$ trials of a two-outcome experiment | $_nC_r p^r (1 - p)^{n-r}$, where $p = $ probability of a success |
| The $k$th term of $(x + y)^n$ | $_nC_{k-1} x^{n-(k-1)} y^{k-1}$ |

# Some Key Math B Facts and Skills

## 1. GEOMETRIC PROOFS

### 1.1 CONGRUENT TRIANGLES

When two triangles are congruent, you can conclude that any pair of corresponding parts are congruent. To *prove* that two triangles are congruent, show that any one of the following conditions is true:

- The three sides of one triangle are congruent to the corresponding parts of the other triangle (SSS ≅ SSS).
- Two sides and their included angle of one triangle are congruent to the corresponding parts of the other triangle (SAS ≅ SAS).
- Two angles and their included side of one triangle are congruent to the corresponding parts of the other triangle (ASA ≅ ASA).
- Two angles and the side opposite one of these angles of one triangle are congruent to the corresponding parts of the other triangle (AAS ≅ AAS).
- The hypotenuse and a leg of one right triangle are congruent to the corresponding parts of the other right triangle (HL ≅ HL).

### 1.2 SIMILAR TRIANGLES

When two figures are similar, you can conclude that any pair of corresponding angles are congruent and that the lengths of corresponding sides are in proportion. To *prove* that two *triangles* are similar, prove that two angles of one triangle are congruent to the corresponding parts of the other triangle.

## 1.3 INDIRECT PROOF

To prove a statement indirectly, assume what needs to be proved is *not* true. Then show that this leads to contradiction of a known fact.

## 1.4 COORDINATE PROOFS

- To prove that a quadrilateral is a parallelogram, use the midpoint formula to show that the diagonals have the same midpoint and, as a result, bisect each other.
- To prove that a quadrilateral is a rhombus, use the distance formula to show that the four sides have the same length.
- To prove that a quadrilateral is a trapezoid, use the slope formula to show that: (1) the slopes of two sides are equal and, as a result, the two sides are parallel; and (2) the slopes of two sides are unequal and, as a result, the two sides are not parallel.
- To prove that two lines are perpendicular, use the slope formula to show that the product of the slopes of the two lines is −1.

# Practice Exercises

**1.1.** Given: circle $O$, $\overline{DB}$ is tangent to the circle at $B$, $\overline{BC}$ and $\overline{BA}$ are chords, and $C$ is the midpoint of $\overset{\frown}{AB}$.

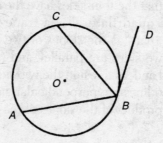

Prove: $\angle ABC \cong \angle CBD$

**1.2.** Given: rectangle $ABCD$, $\overline{BNPC}$, $\overline{AEP}$, $\overline{DEN}$, and $\overline{AP} \cong \overline{DN}$.

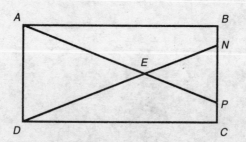

Prove: $a$  $\triangle ABP \cong \triangle DCN$
$\quad\quad\quad b$  $\overline{AE} \cong \overline{DE}$

**1.3.** The vertices of quadrilateral *GAME* are $G(r,s)$, $A(0,0)$, $M(t,0)$, and $E(t + r,s)$. Using coordinate geometry, prove that quadrilateral *GAME* is a parallelogram.

**1.4.** Given: isosceles triangle *ABC*, $\overline{BA} \cong \overline{BC}$, $\overline{AE} \perp \overline{BC}$, and $\overline{BD} \perp \overline{AC}$.

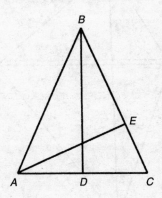

Prove: $\dfrac{AC}{BA} = \dfrac{AE}{BD}$

**1.5.** Given: $\overline{AB} \cong \overline{AC}$
$\overline{BD} \not\cong \overline{CD}$

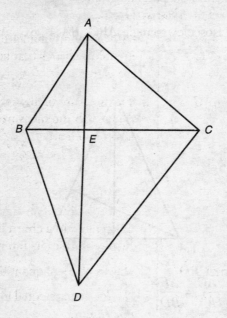

Prove: $\overline{BE} \not\cong \overline{EC}$

## Solutions

**1.1.** Given: Circle $O$, $\overline{DB}$ is tangent to the circle at $B$, $\overline{BC}$ and $\overline{BA}$ are chords, and $C$ is the midpoint of $\overparen{AB}$.

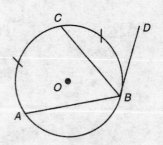

Prove: $\angle ABC \cong \angle CBD$

PLAN: Show that the measures of angles $ABC$ and $CBD$ are one-half the measures of equal arcs, and, as a result, the angles are congruent.

**PROOF**

| Statement | Reason |
|---|---|
| 1. $C$ is the midpoint of $\overset{\frown}{AB}$. | 1. Given. |
| 2. $m\overset{\frown}{AC} = m\overset{\frown}{BC}$ | 2. The midpoint of an arc divides the arc into two arcs that have the same measure. |
| 3. $m\angle ABC = \dfrac{1}{2}m\overset{\frown}{AC}$ | 3. The measure of an inscribed angle is one-half the measure of its intercepted arc. |
| 4. $\overline{DB}$ is tangent to the circle $O$ at $B$. | 4. Given. |
| 5. $m\angle CBD = \dfrac{1}{2}m\overset{\frown}{DC}$ | 5. The measure of an angle formed by a tangent and a chord is one-half the measure of its intercepted arc. |
| 6. $m\angle ABC \cong m\angle CBD$ | 6. Halves of equal quantities are equal. |
| 7. $\angle ABC \cong \angle CBD$ | 7. Angles that are equal in measure are congruent. |

**1.2.** Given: rectangle $ABCD$, $\overline{BNPC}$, $\overline{AEP}$, $\overline{DEN}$, and $\overline{AP} \cong \overline{DN}$.
     Prove: $a$   $\triangle ABP \cong \triangle DCN$
            $b$   $\overline{AE} \cong \overline{DE}$

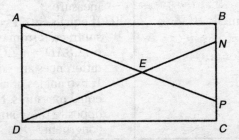

PLAN: **a.** Prove right triangle $ABP \cong$ right triangle $DCN$ by Hyp-Leg $\cong$ Hyp-Leg.

**b.** Prove $\overline{AE} \cong \overline{DE}$ by showing that the angles opposite these sides in $\triangle AED$ are congruent:

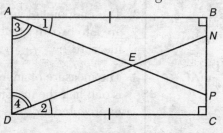

## PROOF

| Statement | Reason |
|---|---|
| **a.** Prove $\triangle ABP \cong \triangle DCN$. | |
| 1. $ABCD$ is a rectangle. | 1. Given. |
| 2. $\angle A$, $\angle B$, $\angle C$, and $\angle D$ are right angles. | 2. A rectangle contains four right angles. |
| 3. $\triangle ABP$ and $\triangle DCN$ are right triangles. | 3. A triangle that contains a right angle is a right triangle. |
| 4. $\overline{AP} \cong \overline{DN}$ (hyp) | 4. Given. |
| 5. $\overline{AB} \cong \overline{CD}$ (leg) | 5. Opposite sides of a rectangle are congruent. |
| 6. $\triangle ABP \cong \triangle DCN$. | 6. Hyp-leg $\cong$ hyp-leg. |
| **b.** Prove $\overline{AE} \cong \overline{DE}$. | |
| 7. $m\angle 1 = m\angle 2$ | 7. Corresponding angles of congruent triangles are equal in measure. |
| 8. $m\angle BAD = m\angle CDA$ | 8. Right angles are equal in measure. |
| 9. $m\angle 3 = m\angle 4$ | 9. If equals ($m\angle 1 = m\angle 2$) are subtracted from equals ($m\angle BAD = m\angle CDA$), the differences are equal. |
| 10. $\overline{AE} \cong \overline{DE}$ | 10. If two angles of a triangle have equal measures, the sides opposite these angles are congruent. |

**1.3.** A quadrilateral is a parallelogram if its diagonals have the same midpoint and, as a result, bisect each other.

- Determine the $x$- and the $y$-coordinates of the midpoint of diagonal $\overline{GM}$ by finding the averages of the corresponding coordinates of $G(r,s)$ and $M(t,0)$:

$$\text{Midpoint of } \overline{GM} = \left(\frac{r+t}{2}, \frac{s+0}{2}\right) = \left(\frac{r+t}{2}, \frac{s}{2}\right)$$

- Determine the $x$- and $y$-coordinates of the midpoint of diagonal $\overline{AE}$ by finding the averages of the corresponding coordinates of $A(0,0)$ and $E(t+r,s)$:

$$\text{Midpoint of } \overline{AE} = \left[\frac{0+(t+r)}{2}, \frac{0+s}{2}\right] = \left(\frac{r+t}{2}, \frac{s}{2}\right)$$

- Diagonals $\overline{GM}$ and $\overline{AE}$ of quadrilateral $GAME$ have the same midpoint; hence they bisect each other.

Therefore, quadrilateral $GAME$ is a parallelogram.

**1.4.** Given: Isosceles triangle $ABC$, $\overline{BA} \cong \overline{BC}$, $\overline{AE} \perp \overline{BC}$, and $\overline{BD} \perp \overline{AC}$.

Prove: $\dfrac{AC}{BA} = \dfrac{AE}{BD}$

PLAN: Show that triangles $AEC$ and $BDA$ are similar and that, as a result, the lengths of their corresponding sides are in proportion.

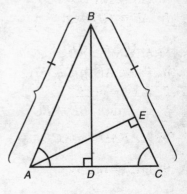

## PROOF

| Statement | Reason |
|---|---|
| 1. $\overline{BA} \cong \overline{BC}$ | 1. Given. |
| 2. $\angle BAD \cong \angle ACE$ (angle) | 2. If two sides of a triangle are congruent, then the angles opposite these sides are congruent. |
| 3. $\overline{AE} \perp \overline{BC}, \overline{BD} \perp \overline{AC}$. | 3. Given. |
| 4. Angles $BDA$ and $AEC$ are right angles | 4. Perpendicular lines meet to form right angles. |
| 5. $\angle BDA \cong \angle AEC$ (angle) | 5. All right angles are congruent. |
| 6. $\triangle AEC \sim \triangle BDA$ | 6. Two triangles are similar if two angles of one triangle are congruent to two angles of the other triangle. |
| 7. $\dfrac{AC}{BA} = \dfrac{AE}{BD}$ | 7. The lengths of corresponding sides of similar triangles are in proportion. |

**1.5.** Given: $\overline{AB} \cong \overline{AC}$

$\overline{BD} \not\cong \overline{CD}$

Prove: $\overline{BE} \not\cong \overline{EC}$

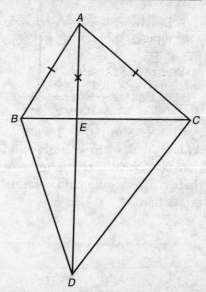

PLAN: Use an indirect proof. Assume that $\overline{BE} \cong \overline{EC}$ and show that this leads to a contradiction of the given fact that $\overline{BD} \not\cong \overline{CD}$.

**PROOF**

| Statement | Reason |
|---|---|
| 1. $\overline{AB} \cong \overline{AC}$ (side) | 1. Given. |
| 2. $\overline{BD} \not\cong \overline{CD}$ | 2. Given. |
| 3. Either $\overline{BE} \not\cong \overline{EC}$ or $\overline{BE} \cong \overline{EC}$ | 3. A statement is either true or false. |
| 4. $\overline{BE} \cong \overline{EC}$ (side) | 4. Assume that the opposite of what needs to be proved is true. |
| 5. $\overline{AE} \cong \overline{AE}$ (side) | 5. Reflexive property of congruence. |
| 6. $\triangle AEB \cong \triangle AEC$ | 6. SSS $\cong$ SSS. |
| 7. $\angle AEB \cong \angle AEC$ | 7. Corresponding parts of congruent triangles are congruent. |
| 8. $\angle BED \cong \angle CED$ (angle) | 8. Supplements of congruent angles are congruent. |
| 9. $\overline{ED} \cong \overline{ED}$ (side) | 9. Reflexive property of congruence. |
| 10. $\triangle BED \cong \triangle CED$ | 10. SAS $\cong$ SAS. |
| 11. $\overline{BD} \cong \overline{CD}$ | 11. Corresponding parts of congruent triangles are congruent. |
| 12. $\overline{BE} \not\cong \overline{EC}$ | 12. Statement 11 contradicts statement 2. The assumption made in statement 4 must be false, and so its opposite must be true. |

# 2. CIRCLES AND ANGLE MEASUREMENT

Angle measurement in circles depends on the location of the vertex of the angle.

## 2.1 VERTEX AT THE CENTER OF THE CIRCLE: CENTRAL ANGLE

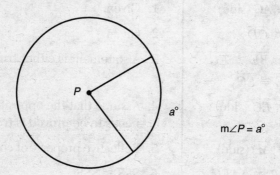

$$m\angle P = a^\circ$$

## 2.2 VERTEX ON THE CIRCLE: INSCRIBED ANGLE OR CHORD-TANGENT ANGLE

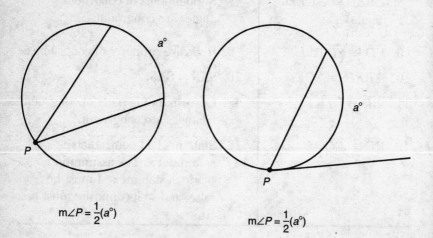

$$m\angle P = \frac{1}{2}(a^\circ) \qquad\qquad m\angle P = \frac{1}{2}(a^\circ)$$

## 2.3 VERTEX INSIDE THE CIRCLE: CHORD-CHORD ANGLE

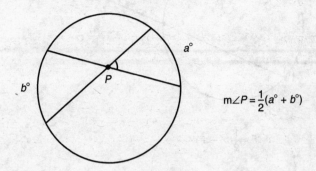

$$m\angle P = \frac{1}{2}(a° + b°)$$

## 2.4 VERTEX OUTSIDE THE CIRCLE: TANGENT-TANGENT, TANGENT-SECANT, AND SECANT-SECANT ANGLES

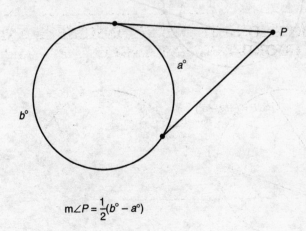

$$m\angle P = \frac{1}{2}(b° - a°)$$

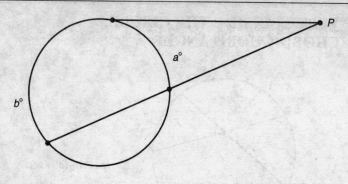

$$m\angle P = \frac{1}{2}(b° - a°)$$

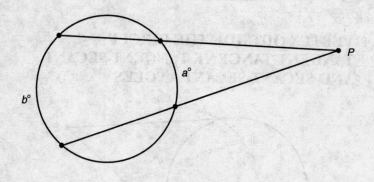

$$m\angle P = \frac{1}{2}(b° - a°)$$

# Practice Exercises

**2.1.** In the accompanying diagram, $\overrightarrow{BD}$ is tangent to circle $O$ at $B$, $\overline{BC}$ is a chord, and $\overline{BOA}$ is a diameter. If $m\widehat{AC}:m\widehat{CB} = 1:4$, find $m\angle DBC$.

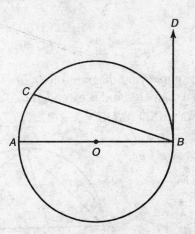

**2.2.** In the diagram below, chords $\overline{AB}$ and $\overline{CD}$ intersect at $E$. If $m\angle AEC = 4x$, $m\widehat{AC} = 120$, and $m\widehat{DB} = 2x$, what is the value of $x$?

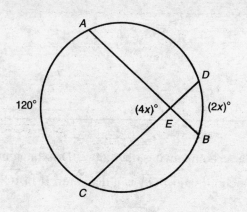

(1) 12
(2) 20

(3) 30
(4) 60

**2.3.** In the accompanying diagram of circle $O$, $\overline{AE}$ and $\overline{FD}$ are chords, $\overline{AOBG}$ is a diameter and is extended to $C$, $\overline{CDE}$ is a secant, $\overline{AE} \parallel \overline{FD}$, and $m\widehat{AE} : m\widehat{ED} : m\widehat{DG} = 5:3:1$.

Find:

*a*   $m\widehat{DG}$
*b*   $m\angle AEF$
*c*   $m\angle DBG$
*d*   $m\angle DCA$
*e*   $m\angle CDF$

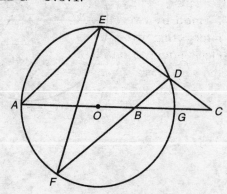

## Solutions

**2.1.** Let $\stackrel{\frown}{mAC} = x$.

Since $m\stackrel{\frown}{AC} : m\stackrel{\frown}{CB} = 1:4$, $m\stackrel{\frown}{CB} = 4x$.

Since $\overline{BOA}$ is a diameter, $\stackrel{\frown}{ACB}$ is a semicircle; the measure of a semicircle is 180:

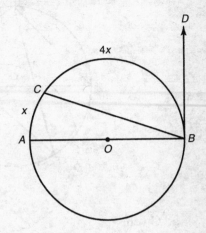

$$m\stackrel{\frown}{ACB} = 180$$
$$x + 4x = 180$$
$$5x = 180$$
$$x = 36$$
$$m\stackrel{\frown}{CB} = 4x = 4(36) = 144$$
$$m\angle DBC = \frac{1}{2}(144) = 72$$

**2.2.** Angle *AEC* is an *angle formed by two chords intersecting within the circle*; the measure of an angle formed by two chords intersecting within a circle is equal to one-half the sum of the measures of the two intercepted arcs:

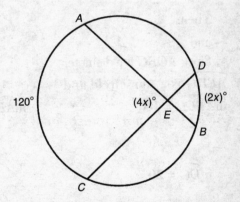

$$m\angle AEC = \frac{1}{2}\left(m\stackrel{\frown}{AC} + m\stackrel{\frown}{DB}\right)$$

$$4x = \frac{1}{2}(120 + 2x)$$

$$4x = 60 + x$$

$$4x - x = 60$$

$$3x = 60$$

$$x = 20$$

The correct choice is **(2)**.

2.3.

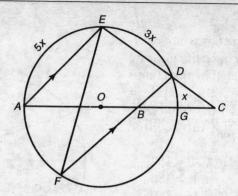

**a.** Given:

Let:

Then:

and

Since $\overline{AOBG}$ is a diameter
$\overset{\frown}{AEDG}$ is a semicircle and
$\overset{\frown}{mAEDG} = 180°$:

$$\text{m}\overset{\frown}{AE} : \text{m}\overset{\frown}{ED} : \text{m}\overset{\frown}{DG} = 5 : 3 : 1$$

$$\text{m}\overset{\frown}{DG} = x$$

$$\text{m}\overset{\frown}{ED} = 3x$$

$$\text{m}\overset{\frown}{AE} = 5x$$

$$\text{m}\overset{\frown}{AE} + \text{m}\overset{\frown}{ED} + \text{m}\overset{\frown}{DG} = \text{m}\overset{\frown}{AEDG}$$
$$5x + 3x + x = 180$$
$$9x = 180$$
$$x = 20$$

$\text{m}\overset{\frown}{DG} = \mathbf{20}$.

**b.** From part **a**, $\text{m}\overset{\frown}{ED} = 3x$. $\overline{AE} \parallel \overline{FD}$
and parallel chords intercept on a circle
arcs that are equal in measure:

From part **a**, $x = 20$:

$$\text{m}\overset{\frown}{AF} = \text{m}\overset{\frown}{ED} = 3x$$
$$\text{m}\overset{\frown}{AF} = 3(20) = 60$$

Angle $AEF$ is an *inscribed angle*; the
measure of an inscribed angle is equal to
one-half the measure of its intercepted
arc:

$$\text{m}\angle AEF = \frac{1}{2}\text{m}\overset{\frown}{AF}$$

$$= \frac{1}{2}(60) = 30$$

$\text{m}\angle AEF = \mathbf{30}$.

**c.** Angle *DBG* is an *angle formed by two chords intersecting within the circle*; the measure of an angle formed by two chords intersecting within a circle is equal to one-half the sum of the measures of the intercepted arcs:

From part **a**, $\stackrel{\frown}{mDG} = 20$; from part **b**, $\stackrel{\frown}{mAF} = 60$:

$$m\angle DBG = \frac{1}{2}\left(m\stackrel{\frown}{AF} + m\stackrel{\frown}{DG}\right)$$

$$m\angle DBG = \frac{1}{2}(60 + 20)$$

$$= \frac{1}{2}(80) = 40$$

$m\angle DBG = \mathbf{40}$.

**d.** Angle *DCA* is an *angle formed by two secants intersecting outside the circle*; the measure of an angle formed by two secants intersecting outside a circle is equal to one-half the difference of the measures of the intercepted arcs:

From part **a**, $m\stackrel{\frown}{DG} = 20$; also $m\stackrel{\frown}{AE} = 5x$ *and* $x = 20$; therefore, $m\stackrel{\frown}{AE} = 5(20) = 100$:

$$m\angle DCA = \frac{1}{2}\left(m\stackrel{\frown}{AE} - m\stackrel{\frown}{DC}\right)$$

$$m\angle DCA = \frac{1}{2}(100 - 20)$$

$$= \frac{1}{2}(80) = 40$$

$m\angle DCA = \mathbf{40}$.

**e.** Angle *EDF* is an *inscribed angle*; the measure of an inscribed angle is equal to one-half the measure of its intercepted arc:

$$m\angle EDF = \frac{1}{2}m\stackrel{\frown}{FAE}$$

$$m\stackrel{\frown}{FAE} = m\stackrel{\frown}{AF} + m\stackrel{\frown}{AE}$$

From part **b**, $\text{m}\overset{\frown}{AF} = 60$;

from part **d**, $\text{m}\overset{\frown}{AE} = 100$:

$$\text{m}\overset{\frown}{FAE} = 60 + 100 = 160$$

$$\text{m}\angle EDF = \frac{1}{2}(160) = 80$$

Angle $CDF$ is the supplement of $\angle EDF$; the sum of the measures of two supplementary angles is 180°:

$$\text{m}\angle CDF + \text{m}\angle EDF = 180$$
$$\text{m}\angle CDF + 80 = 180$$
$$\text{m}\angle CDF = 180 - 80 = 100$$

$\text{m}\angle CDF = \mathbf{100}$.

# 3. INTERSECTING CHORDS; TANGENTS AND SECANTS

## 3.1 INTERSECTING CHORDS

When two chords intersect inside a circle, the product of the lengths of the segments of one chord is equal to the product of the lengths of the segments of the other chord.

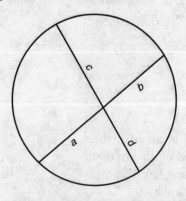

$$a \times b = c \times d$$

## 3.2 TANGENT-SECANT SEGMENTS

When a tangent and a secant intersect outside a circle, the square of the length of the tangent is equal to the product of the lengths of the secant and the part of the secant outside the circle.

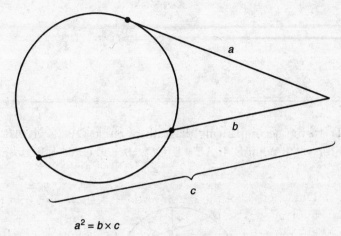

$$a^2 = b \times c$$

## 3.3 SECANT-SECANT SEGMENTS

When two secants intersect outside a circle, the product of the lengths of the first secant and its external segment is equal to the product of the lengths of the second secant and its external segment.

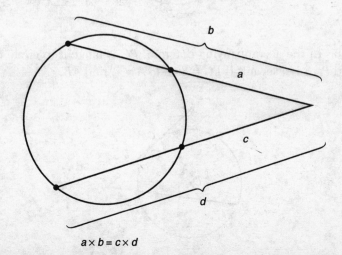

$$a \times b = c \times d$$

# Practice Exercises

**3.1.** In the accompanying diagram of circle $O$, chords $\overline{AB}$ and $\overline{CD}$ intersect at point $E$. If $AE = 2$, $CD = 9$, and $CE = 4$, find $BE$.

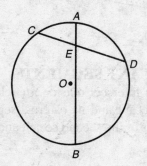

**3.2.** In the accompanying diagram, $\overrightarrow{PC}$ is tangent to circle $O$ at $C$ and $\overline{PAB}$ is a secant. If $PC = 8$ and $PA = 4$, find $AB$.

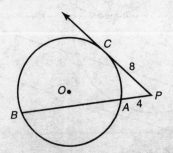

**3.3.** In the accompanying diagram of circle $O$, secants $\overline{CBA}$ and $\overline{CED}$ intersect at $C$. If $AC = 12$, $BC = 3$, and $DC = 9$, find $EC$.

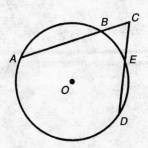

**3.4.** In the accompanying diagram, $\overline{AFB}$, $\overline{AEC}$, and $\overline{BGC}$ are tangent to circle $O$ at $F$, $E$, and $G$, respectively. If $AB = 32$, $AE = 20$, and $EC = 24$, find $BC$.

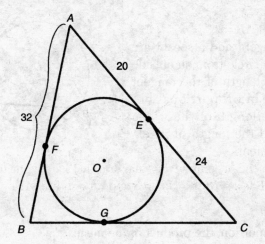

## Solutions

**3.1.** Let $BE = x$.

If two chords intersect within a circle, the product of the lengths of the segments of one chord equals the product of the lengths of the segments of the other chord:

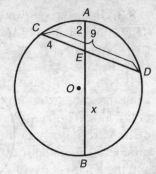

$$DE = CD - CE = 9 - 4 = 5.$$

Therefore:

$$AE \times BE = CE \times ED$$
$$2x = 4(5)$$
$$2x = 20$$
$$x = 10$$

$BE = \mathbf{10}$.

**3.2.** Let $AB = x$.

If a tangent and a secant are drawn to a circle from an outside point, the length of the tangent is the mean proportional between the length of the whole secant and the length of its external segment:

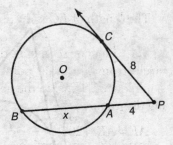

$$\frac{PB}{PC} = \frac{PC}{PA}$$
$$\frac{x+4}{8} = \frac{8}{4}$$

In a proportion, the product of the means equals the product of the extremes (cross-multiply):

$$4(x + 4) = 8(8)$$
$$4x + 16 = 64$$
$$4x = 64 - 16$$
$$4x = 48$$
$$x = 12$$

$AB = \mathbf{12}$.

**3.3.** It is given that, in the accompanying diagram of circle $O$, secants $\overline{CBA}$ and $\overline{CED}$ intersect at $C$, $AC = 12$, $BC = 3$, and $DC = 9$.

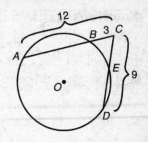

If two secants are drawn to a circle from the same exterior point, the product of the length of one secant and the length of its external segment is equal to the product of the length of the other secant and the length of its external segment.

Thus:

Since $AC = 12$, $BC = 3$, and $DC = 9$:

$$AC \times BC = DC \times EC$$
$$12 \times 3 = 9 \times EC$$
$$36 = 9(EC)$$
$$\frac{36}{9} = EC$$
$$4 = EC$$

The length of $\overline{EC}$ is **4**.

**3.4.** Tangent segments drawn to a circle from the same point have the same length. Thus, in the accompanying diagram:

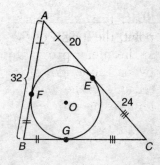

$$AF = AE = 20$$
$$BF = 32 - AF = 32 - 20 = 12$$
$$BG = BF = 12$$
$$CG = CE = 24$$
$$BC = BG + CG = 12 + 24 = 36$$

$$BC = \mathbf{36}.$$

# 4. PROPORTIONS IN RIGHT TRIANGLES

If the altitude is drawn to the hypotenuse of a right triangle, then each of the right triangles formed is similar to the original right triangle. Furthermore, the following proportions are true:

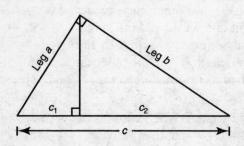

$$\frac{\text{hyp segment } c_1}{\text{altitude}} = \frac{\text{altitude}}{\text{hyp segment } c_2}$$

$$\frac{\text{hyp } c}{\text{leg } a} = \frac{\text{leg } a}{\text{hyp segment } c_1}$$

$$\frac{\text{hyp } c}{\text{leg } b} = \frac{\text{leg } b}{\text{hyp segment } c_2}$$

# Practice Exercises

**4.1.** In right triangle $ABC$, altitude $\overline{CD}$ is drawn to hypotenuse $\overline{AB}$. If $AD = 2$ and $DB = 6$, then $AC$ is

    (1) $4\sqrt{3}$                       (3) $3$

    (2) $2\sqrt{3}$                       (4) $4$

**4.2.** In right triangle $ABC$, altitude $\overline{CD}$ is drawn to hypotenuse $\overline{AB}$. If $AD = 5$ and $DB = 24$, what is the length of $\overline{CD}$?

    (1) $120$                       (3) $2\sqrt{30}$

    (2) $\sqrt{30}$                     (4) $4\sqrt{30}$

**4.3.** In $\triangle ABC$, $\overline{AB} \perp \overline{BC}$ and $\overline{DE} \perp \overline{CA}$. If $DE = 8$, $CD = 10$, and $CA = 30$, find $AB$.

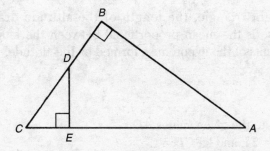

## Solutions

**4.1.** If an altitude is drawn to the hypotenuse of a right triangle, the length of either leg is the mean proportional between the hypotenuse and the segment of the hypotenuse that is adjacent to that leg. Thus, since $AD = 2$ and $DB = 6$:

$$\frac{AB}{AC} = \frac{AC}{AD}$$

$$\frac{2+6}{AC} = \frac{AC}{2}$$

$$\frac{8}{AC} = \frac{AC}{2}$$

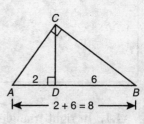

Cross-multiply:    $(AC)^2 = 16$

$$AC = \sqrt{16} = 4$$

The correct choice is **(4)**.

**4.2.**

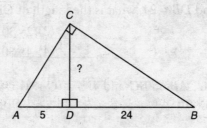

In a right triangle, the length of the altitude drawn to the hypotenuse is the mean proportional between the lengths of the two segments of the hypotenuse formed by the altitude:

$$\frac{AD}{CD} = \frac{CD}{DB}$$

Substitute the given values $AD = 5$ and $DB = 24$, and let $CD = x$:

In a proportion the product of the means equals the product of the extremes (cross-multiply):

$$\frac{5}{x} = \frac{x}{24}$$

$$x \cdot x = 5 \cdot 24$$

$$x^2 = 120$$

Take the square root of each side of the equation:

$$x = \pm\sqrt{120}$$

Reject the negative value of $x$ since $x$ represents the length of a side of the triangle:

$$x = \sqrt{120}$$

Factor the radicand so that one of its factors is the highest perfect square factor of 120:

$$x = \sqrt{4 \cdot 30}$$

Distribute the radical sign over each factor of the radicand:

$$x = \sqrt{4} \cdot \sqrt{30}$$

Evaluate $\sqrt{4}$:

$$= 2\sqrt{30}$$

The correct choice is **(3)**.

**4.3.**

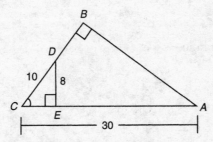

Consider right triangles $DEC$ and $ABC$. Since $\angle E \cong B$ (right angles are congruent) and $\angle C \cong \angle C$ (reflexive property of congruence), these triangles are similar by the $AA \cong AA$ theorem of similarity.

The lengths of corresponding sides of similar triangles are in proportion:

$$\frac{\text{side in } \triangle ABC}{\text{Corresponding side in } \triangle DEC} = \frac{AB}{DE} = \frac{CA}{CD}$$

$$\frac{AB}{8} = \frac{30}{10}$$

In a proportion, the product of the means equals the product of the extremes (cross-multiply):

$$10(AB) = 8(30)$$

$$AB = \frac{240}{10} = 24$$

The length of $\overline{AB}$ is **24**.

# 5. ALGEBRAIC OPERATIONS

A wide variety of algebraic operations is covered in Math B as illustrated in the accompanying set of practice exercises.

# Practice Exercises

**5.1.** Factor completely: $x^3 - x^2 - 6x$

**5.2.** If $x = 5^a$, then the value of $5x$ is
(1) $x + 1$                                  (3) $a + 5$
(2) $6^a$                                    (4) $5^{a+1}$

**5.3.** If $h \neq 0$, when the fraction $\dfrac{(x+h)^2 - x^2}{h}$ is simplified, the result is
(1) $h$                                   (3) $2x^2 + 2x + h$
(2) $0$                                   (4) $2x + h$

**5.4.** Express in simplest form: $\dfrac{x - \dfrac{4}{x}}{1 + \dfrac{2}{x}}$

**5.5.** Combine and express in simplest form:
$$\frac{y - 20}{y^2 - 16} + \frac{2}{y - 4}$$

**5.6.** Express in simplest form:
$$\frac{x^2 - 3x}{x^2 + 3x - 10} \cdot \frac{2x + 10}{3} \div \frac{x^2 - x - 6}{x^2 - 4}$$

**5.7.** The expression $(2 + 3i)^2$ is equal to
(1) $-5$            (3) $-5 + 12i$
(2) $13$            (4) $13 + 12i$

**5.8.** Express $\dfrac{4}{3+\sqrt{2}}$ as an equivalent fraction with a rational denominator.

**5.9.** Evaluate $x^{3/4} - x^0$ if $x = 16$.

**5.10.** The solution set of the equation $\sqrt{x + 1} + 5 = 0$ is
(1) $\phi$            (3) $\{-26\}$
(2) $\{24\}$            (4) $\{0\}$

**5.11.** The solution set of $|x - 2| < 3$ is
(1) $\{x|x > 5\}$            (3) $\{x|-1 < x < 5\}$
(2) $\{x|x < -1\}$            (4) $\{x|x < -1 \text{ or } x > 5\}$

**5.12.** Which is the solution set for the inequality $|2x - 1| < 3$?

(1)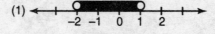

(2)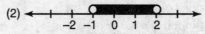

(3)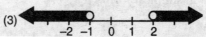

(4)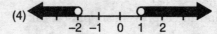

## Solutions

**5.1.** The given trinomial is:
$$x^3 - x^2 - 6x$$

The trinomial contains a *highest common monomial factor* of $x$. The other factor is obtained by dividing each term by $x$:
$$x(x^2 - x - 6)$$

The expression $x^2 - x - 6$ is a *quadratic trinomial* that can be factored into the product of two binomials. Be sure to check that the product of the inner terms added to the product of the outer terms of the binomials equals the middle term, $-x$, of the quadratic trinomial:

$$-3x = \text{inner product}$$
$$x(x - \overset{\frown}{3)(x} + 2)$$
$$+2x = \text{outer product}$$

Since $(-3x) + (+2x) = -x$, these are the correct factors:
$$x(x - 3)(x + 2)$$

The factored form is $\boldsymbol{x(x - 3)(x + 2)}$.

**5.2.** Given:
$$x = 5^a$$
Multiply both sides of the equation by 5:
$$5x = 5(5^a)$$
Both $5^a$ and 5 (which is $5^1$) are powers of 5. Powers of the same base are multiplied by adding their exponents:
$$= 5^{a+1}$$

The correct choice is **(4)**.

**5.3.** The given fraction is:
$$\frac{(x+h)^2 - x^2}{h}$$

Square $(x + h)$:

$$
\begin{array}{r}
x \;+\; h \\
\underline{x \;+\; h} \\
x^2 +\; hx \\
\underline{\phantom{x^2 +} hx + h^2} \\
x^2 + 2hx + h^2
\end{array}
$$

$$\frac{x^2 + 2hx + h^2 - x^2}{h}$$

Combine like terms in the numerator:

$$\frac{2hx+h^2}{h}$$

Factor out the common monomial factor in the numerator:

$$\frac{h(2x+h)}{h}$$

Divide numerator and denominator by $h$; it is given that $h \neq 0$, so division is possible:

$$\frac{\overset{1}{\cancel{h}}(2x+h)}{\underset{1}{\cancel{h}}}$$

$$2x+h$$

The correct choice is **(4)**.

**5.4.** The given expression is a *complex fraction*:

$$\frac{x-\dfrac{4}{x}}{1+\dfrac{2}{x}}$$

Multiply the expression by 1 in the form $\dfrac{x}{x}$:

$$\frac{x\left(x-\dfrac{4}{x}\right)}{x\left(1+\dfrac{2}{x}\right)}$$

$$\frac{x^2-4}{x+2}$$

The numerator can be factored as the difference of two perfect squares:

$$\frac{(x+2)(x-2)}{x+2}$$

Cancel the factor $x+2$, which appears in both numerator and denominator:

$$\frac{\overset{1}{\cancel{(x+2)}}(x-2)}{\underset{1}{\cancel{x+2}}}$$

The result in simplest form is **$x-2$**.

$$x-2$$

**5.5.** The given expression is:

$$\frac{y-20}{y^2-16}+\frac{2}{y-4}$$

The denominator of the first fraction can be factored as the difference of two squares:

$$\frac{y-20}{(y+4)(y-4)}+\frac{2}{y-4}$$

Find the least common denominator (LCD). The LCD is the simplest expression into which each of the denominators will divide evenly. The LCD for $(y-4)$ and $(y+4)(y-4)$ is $(y+4)(y-4)$. Convert the second fraction into an equivalent fraction having the LCD by multiplying it by 1 in the form $\dfrac{y+4}{y+4}$ :

$$\frac{y-20}{(y+4)(y-4)}+\frac{2(y+4)}{(y+4)(y-4)}$$

If fractions have the same denominator, they can be combined by combining their numerators:

$$\frac{y-20+2(y+4)}{(y+4)(y-4)}$$

$$\frac{y-20+2y+8}{(y+4)(y-4)}$$

$$\frac{3y-12}{(y+4)(y-4)}$$

Factor out the common factor, 3, in the numerator:

$$\frac{3(y-4)}{(y+4)(y-4)}$$

Divide the numerator and denominator by the factor $(y-4)$:

$$\frac{\overset{1}{3\cancel{(y-4)}}}{(y+4)\cancel{(y-4)}}$$
$$\underset{1}{}$$

$$\frac{3}{y+4}$$

The expression in simplest form is $\dfrac{3}{y+4}$.

**5.6.** The given expression is:

$$\frac{x^2 - 3x}{x^2 + 3x - 10} \cdot \frac{2x + 10}{3} \div \frac{x^2 - x - 6}{x^2 - 4}$$

Division by the last fraction is equivalent to multiplication by that fraction inverted:

$$\frac{x^2 - 3x}{x^2 + 3x - 10} \cdot \frac{2x + 10}{3} \cdot \frac{x^2 - 4}{x^2 - x - 6}$$

Factor each numerator and denominator where possible:

$$\frac{x(x - 3)}{(x + 5)(x - 2)} \cdot \frac{2(x + 5)}{3} \cdot \frac{(x - 2)(x + 2)}{(x - 3)(x + 2)}$$

Divide out any factor that appears in both a numerator and a denominator:

$$\frac{x\cancel{(x-3)}}{\cancel{(x+5)}\cancel{(x-2)}} \cdot \frac{2\cancel{(x+5)}}{3} \cdot \frac{\cancel{(x-2)}\cancel{(x+2)}}{\cancel{(x-3)}\cancel{(x+2)}}$$

$$\frac{x \cdot 2}{3} \text{ or } \frac{2x}{3}$$

The expression in simplest form is $\dfrac{2x}{3}$.

**5.7.** Given: $(2 + 3i)^2$

Multiply out $(2 + 3i)(2 + 3i)$:

$$\begin{array}{r} 2 + 3i \\ 2 + 3i \\ \hline 4 + 6i \\ + 6i + 9i^2 \\ \hline 4 + 12i + 9i^2 \end{array}$$

Since $i = \sqrt{-1}$, $i^2 = -1$:

$$4 + 12i + 9i^2$$
$$4 + 12i + 9(-1)$$
$$4 + 12i - 9$$
$$-5 + 12i$$

The correct choice is **(3)**.

**5.8.** The given fraction has an irrational denominator:

$$\frac{4}{3+\sqrt{2}}$$

To rationalize the denominator, first determine its *conjugate*. If an expression is of the form $A + \sqrt{B}$, its conjugate is of the form $A - \sqrt{B}$. Thus, the conjugate of $3 + \sqrt{2}$ is $3 - \sqrt{2}$. Multiply the given fraction by 1 in the form $\dfrac{3-\sqrt{2}}{3-\sqrt{2}}$:

$$\frac{(3-\sqrt{2})(4)}{(3-\sqrt{2})(3+\sqrt{2})}$$

The denominator is now of the form $(A - B)(A + B)$ with $A = 3$ and $B = \sqrt{2}$. The product of $(A - B)(A + B)$ is $A^2 - B^2$:

$$\frac{4(3-\sqrt{2})}{(3)^2 - (\sqrt{2})^2}$$

$$\frac{4(3-\sqrt{2})}{9-2}$$

$$\frac{4(3-\sqrt{2})}{7}$$

The equivalent fraction with a rational denominator is $\dfrac{4(3-\sqrt{2})}{7}$.

**5.9.** The given expression is:

$$x^{3/4} - x^0$$

To evaluate it for $x = 16$, substitute 16 for $x$:

$$16^4 - 16^0$$

$x^{m/n} = \sqrt[n]{x^m}$ or $\left(\sqrt[n]{x}\right)^m$, and $x^0 = 1$ if $x \neq 0$:

$$\left(\sqrt[4]{16}\right)^3 - 1$$

$\sqrt[4]{16} = 2$:

$$2^3 - 1$$

$$8-1$$

$$7$$

The value is **7**.

**5.10.** The given equation is a *radical equation*:

$$\sqrt{x+1}+5=0$$

Isolate the radical on one side of the equation:

$$\sqrt{x+1}=-5$$

Square both sides of the equation:

$$x+1=25$$

$$x=25-1$$

$$=24$$

Squaring both sides of an equation may possibly introduce an *extraneous root*. Therefore, the supposed root, 24, must be checked by substituting it in the *original* equation to see whether the equation is satisfied:

$$\sqrt{24+1}+5 \overset{?}{=} 0$$

$$\sqrt{25}+5 \overset{?}{=} 0$$

$$5+5 \overset{?}{=} 0$$

$$10 \neq 0 \quad \text{24 is an extraneous root.}$$

The equation has no root. Its solution set is the null set, denoted by $\emptyset$.

The correct choice is (**1**).


**5.11.** The given inequality contains the absolute value of one expression:

$$|x-2|<3$$

$|n|$ stands for the *absolute value* of $n$; $|n|=n$ if $n \geq 0$, but $|n|=-n$ if $n<0$. In other words, $|n|$ is never negative.

We do not know whether $(x-2)$ is positive or negative. Therefore:

$$
\begin{array}{lll}
\text{Either} & x-2<3 & \text{or} \quad -(x-2)<3 \\
& x<3+2 & \quad\quad -x+2<3 \\
& x<5 & \quad\quad 2-3<x \\
& & \quad\quad -1<x \\
& & \quad (\text{or } x>-1)
\end{array}
$$

Combining the two results, we obtain the solution set:

$$\{x|-1<x<5\}.$$

The correct choice is (**3**).

**5.12.** The symbol $|n|$ represents the *absolute value* of $n$. If $n \geq 0$, $|n| = n$, but if $n < 0$, $|n| = -n$.

The given inequality is:                              $|2x - 1| < 3$

Since we do not know whether $(2x - 1)$ is positive or negative:

Either    $2x - 1 < 3$      or      $-(2x - 1) < 3$

               $2x < 3 + 1$                   $2x - 1 > -3$

               $2x < 4$                         $2x > -3 + 1$

               $x < 2$                         $2x > -2$

                                                  $x > -1$

Note that, in solving the right-hand inequality, both sides are multiplied by $-1$. When two sides of an inequality are multiplied by a negative number, the direction of the inequality must be reversed.

The solution to the inequality is $-1 < x < 2$.

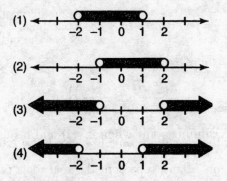

Consider each choice in turn:

(1) represents $-2 < x < 1$. The open circles at $-2$ and $1$ indicate that $-2$ and $1$ are not included in the set.

(2) represents $-1 < x < 2$. This is the correct choice.

(3) represents $x < -1$ or $x > 2$.

(4) represents $x < -2$ or $x > 1$.

The correct choice is **(2)**.

# 6. QUADRATIC EQUATIONS AND INEQUALITIES

## 6.1 QUADRATIC FORMULA
If $ax^2 + bx + c = 0$, where $a \neq 0$, then:

- $x = \dfrac{-b \pm \sqrt{b^2 - 4ac}}{2a}$.

- Sum of the roots $= -\dfrac{b}{a}$, and product of the roots $= \dfrac{c}{a}$.

- The discriminant, $b^2 - 4ac$, determines the nature of the roots. Roots are rational when $b^2 - 4ac$ is a perfect square; roots are irrational when $b^2 - 4ac$ is positive and not a perfect square; roots are equal when $b^2 - 4ac = 0$; and roots are imaginary when $b^2 - 4ac < 0$.

## 6.2 PARABOLA
The graph of $y = f(x) = ax^2 + bx + c$ is a parabola.

- The vertex (turning point) is at

$$\left( -\frac{b}{2a}, f\left( -\frac{b}{2a} \right) \right)$$

which is a maximum point when $a < 0$ and a minimum point when $a > 0$.
- The $x$-intercepts, if any, represent the real roots of $ax^2 + bx + c = 0$.

## 6.3 QUADRATIC INEQUALITIES
If $r_1$ and $r_2$ are the roots of $ax^2 + bx + c = 0$, where $a > 0$ and $r_1 < r_2$, then:

- $r_1 < x < r_2$ is the solution set of $ax^2 + bx + c < 0$.
- $x < r_1$ or $x > r_2$ is the solution set of $ax^2 + bx + c > 0$.

# Practice Exercises

**6.1.** If the discriminant of an equation is 10, then the roots are
(1) real, rational, and unequal
(2) real, irrational, and unequal
(3) real, rational, and equal
(4) imaginary

**6.2.** What is the product of the roots of the equation $-2x^2 + 3x + 8 = 0$?
(1) $\dfrac{3}{2}$  (3) $\dfrac{3}{4}$
(2) $-4$  (4) $4$

**6.3.** Solve the equation $2(x - 3) = -\dfrac{5}{x}$ and express its roots in terms of $i$.

**6.4.** What is the solution set for $x^2 - x - 6 < 0$?
(1) $\{x | x < -2 \text{ or } x > 3\}$  (3) $\{x | -3 < x < 2\}$
(2) $\{x | x < -3 \text{ or } x > 2\}$  (4) $\{x | -2 < x < 3\}$

## Solutions

**6.1.** The *discriminant* of a quadratic equation of the form $ax^2 + bx + c = 0$ is the value of $b^2 - 4ac$.

The roots of the quadratic equation are

$$\frac{-b \pm \sqrt{b^2 - 4ac}}{2a}$$

If $b^2 - 4ac = 10$, then the roots are in the form $\dfrac{-b \pm \sqrt{10}}{2a}$; that is, they are real, irrational, and unequal.

The correct choice is (**2**).

**6.2.** If a quadratic equation is in the form $ax^2 + bx + c = 0$, then the product of its roots is $\dfrac{c}{a}$.

The given equation $-2x^2 + 3x + 8 = 0$ is in the form $ax^2 + bx + c = 0$, with $a = -2$, $b = 3$, and $c = 8$. The product of its root is $\dfrac{8}{-2}$ or $-4$.

The correct choice is (**2**).

**6.3.** The given equation is:
$$2(x - 3) = -\frac{5}{x}$$

Remove the parentheses by applying the Distributive Law:
$$2x - 6 = -\frac{5}{x}$$

Clear fractions by multiplying each term on both sides of the equation by $x$:
$$2x^2 - 6x = -5$$

This is a *quadratic equation*; rearrange it so that all terms are on one side equal to 0:
$$2x^2 - 6x + 5 = 0$$

The left side is a quadratic trinomial that cannot be factored, so the equation must be solved by using the *quadratic formula*.

If a quadratic equation is in the form $ax^2 + bx + c = 0$, then:
$$x = \frac{-b \pm \sqrt{b^2 - 4ac}}{2a}$$

The equation $2x^2 - 6x + 5 = 0$ is in the form $ax^2 + bx + c = 0$, with $a = 2$, $b = -6$, and $c = 5$:
$$x = \frac{-(-6) \pm \sqrt{(-6)^2 - 4(2)(5)}}{2(2)}$$

$$x = \frac{6 \pm \sqrt{36 - 40}}{4}$$

$$x = \frac{6 \pm \sqrt{-4}}{4}$$

$$x = \frac{6 \pm \sqrt{4}\sqrt{-1}}{4}$$

Since $\sqrt{4} = 2$ and $\sqrt{-1} = i$:              $x = \dfrac{6 \pm 2i}{4}$

Reduce the fraction by dividing all terms in the numerator and denominator by 2:              $x = \dfrac{3 \pm i}{2}$

The roots are $\dfrac{3 + i}{2}$.

**6.4.** The given inequality is:              $x^2 - x - 6 < 0$

The left side is a *quadratic trinomial* that can be factored into the product of two binomials. Be sure to check that the sum of the inner and outer cross-products of the binomials equals the middle term, $-x$, of the original trinomial:      $-3x = $ inner product

$$(x - \overgroup{3)(x} + 2) < 0$$

$+2x = $ outer product

Since $(-3x) + (+2x) = -x$, these are the correct factors:             $(x - 3)(x + 2) < 0$

The inequality shows that the product of two factors is less than zero, that is, is negative. Therefore, one factor must be positive and the other must be negative. There are two possibilities:

$$(x - 3 < 0 \text{ AND } x + 2 > 0) \quad \text{or} \quad (x - 3 > 0 \text{ AND } x + 2 < 0)$$
$$(x < 3 \text{ AND } x > -2) \quad \text{or} \quad (x > 3 \text{ AND } x < -2)$$

But $x > 3$ and $x < -2$ are contradictory, so this possibility must be ruled out.

The only solution is:       $x < 3$ and $x > -2$ or $-2 < x < 3$

The solution set of the inequality is:           $\{x \mid -2 < x < 3\}$

The correct choice is (**4**).

# 7. FUNCTIONS AND TRANSFORMATIONS

## 7.1 FUNCTION AND INVERSE FUNCTION

A function is a set of ordered pairs in which no two ordered pairs have the same $x$-value but different $y$-values.

- An equation represents a function if no vertical line intersects its graph in more than one point.
- A function has an inverse function if no horizontal line intersects its graph in more than one point.
- If a function has an inverse, then it can be formed by interchanging $x$ and $y$ and then solving for $y$ in terms of $x$.

## 7.2 REFLECTION AND TRANSLATION RULES

- $r_{x\text{-axis}}(x,y) = (x,-y)$.
- $r_{y\text{-axis}}(x,y) = (-x,y)$.
- $r_{\text{origin}}(x,y) = (-x,-y)$.
- $r_{y=x}(x,y) = (y,x)$.
- $T_{h,k}(x,y) \to (x+h, y+k)$.

## 7.3 ROTATION AND DILATION RULES

- $R_{90°}(x,y) = (-y,x)$.
- $R_{180°}(x,y) = (-x,-y)$.
- $R_{270°}(x,y) = (y,-x)$.
- $D_k(x,y) = (kx,ky)$, where $k \neq 0$.

# Practice Exercises

**7.1.** Write an equation of the line of reflection that maps $A(1,5)$ onto $A'(5,1)$.

**7.2.** Which transformation is *not* an isometry?
    (1) $T_{(5,3)}$                           (3) $r_{x\text{-axis}}$
    (2) $D_2$                               (4) $R_{(0,90°)}$

**7.3.** What is the image of $P(-4,6)$ under the composite $r_{x=2} \circ r_{y\text{-axis}}$?
    (1) $(-8,6)$                       (3) $(6,0)$
    (2) $(4,-2)$                      (4) $(0,6)$

**7.4.** What is the domain of the function $f(x) = \dfrac{4}{\sqrt{x+1}}$ over the set of real numbers?
    (1) $\{x | x = 1\}$                  (3) $\{x | x < -1\}$
    (2) $\{x | x \geq -1\}$               (4) $\{x | x > -1\}$

**7.5.** If $f(x) = kx^2$ and $f(2) = 12$, then $k$ equals
    (1) 1                                  (3) 3
    (2) 2                                  (4) 4

**7.6.** If $f(x) = \dfrac{x-4}{x+4}$, then $f(4a)$ equals
    (1) $\dfrac{a-1}{a+1}$                  (3) $\dfrac{4a-1}{4a+1}$
    (2) $\dfrac{a+1}{a-1}$                  (4) $\dfrac{4a+1}{4a-1}$

**7.7.** Given: set $A = \{(1,2),(2,3),(3,4),(4,5)\}$
If the inverse of the set is $A^{-1}$, which statement is true?
(1) $A$ and $A^{-1}$ are functions.
(2) $A$ and $A^{-1}$ are not functions.
(3) $A$ is a function and $A^{-1}$ is not a function.
(4) $A$ is not a function and $A^{-1}$ is a function.

**7.8.** If $f(x) = x - 3$ and $g(x) = x^2$, what is the value of $(f \circ g)(2)$?

**7.9.** Given: $J = -2 + 5i$ and $K = 3 + 2i$
*a* On graph paper, plot and label $J$ and $K$.
*b* On the same set of axes, plot the sum of $J$ and $K$ and label it $L$.
*c* On the same set of axes, plot the image of $L$ after a counterclockwise rotation of $270°$ and label it $L'$.
*d* Express $L'$ as a complex number.

## Solutions

**7.1.** If $A(1,5)$ is mapped onto $A'(5,1)$, this mapping replaces $x$ by $y$ and $y$ by $x$:

$$A(x,y) \rightarrow A'(y,x)$$

Such a mapping is produced by a reflection in a line through the origin inclined at $45°$ to the positive directions of both the $x$- and $y$-axes.

An equation of this line of reflection is $y = x$.

The line of reflection is $y = x$.

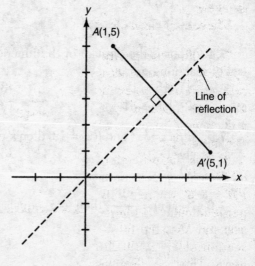

**7.2.** An *isometry* is a transformation that results in the mapping of a figure onto an image that is congruent to the original figure.

Consider each choice in turn:

(1) $T_{(5,3)}$ is a *translation* which maps a point $P_{(x,y)}$ onto its image, $P'(x + 5, y + 3)$. Since all points are moved 5 units to the right and 3 units up, the distance between any two points is the same as the distance between their respective images. Thus, $T_{(5,3)}$ is an isometry.

(2) $D_2$ is a *dilation* which maps a point $P(x,y)$ onto its image, $P'(2x,2y)$. The distance between the images of any two points becomes double the distance between the original two points. Thus, $D_2$ is not an isometry.

(3) $r_{x\text{-axis}}$ is a *reflection* in the x-axis which maps a point $P(x,y)$ onto its image, $P'(x,-y)$. The result is equivalent to "flipping" figures over the x-axis; all mirror images remain congruent to the original figures. Thus, $r_{x\text{-axis}}$ is an isometry.

(4) $R_{(0,90°)}$ is a *rotation* of 90° about the origin. All original figures remain congruent to their images; they are simply rotated $\frac{1}{4}$ of a turn to assume the positions of their images. Thus, $R_{(0,90°)}$ is an isometry.

The correct choice is (**2**).

**7.3.** The composite $r_x = 2 \circ r_{y\text{-axis}}$ represents a reflection of a point in the y-axis followed by a reflection of its image in the line $x = 2$.

If $P'$ is the reflection of $P$ in the y-axis, then $PM = P'M$, where $\overline{PP'}$ is perpendicular to the y-axis and $M$ is the intersection of $\overline{PP'}$ with the y-axis. Thus, a reflection in the y-axis replaces a

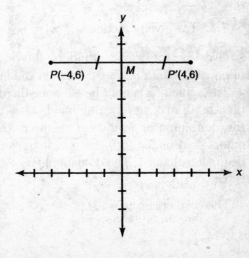

point $P(x,y)$ by its image $P'(-x,y)$. Therefore, $r_{y\text{-axis}} P(-4, 6) \rightarrow P'(4,6)$.

$P'(4,6)$ is next re-flected in the line $x = 2$. The line $x = 2$ is a vertical line 2 units to the right of the $y$-axis. If $P''$ is the reflection of $P'$ in the line $x = 2$, then $\overline{P'N} = \overline{P''N}$, where $\overline{P'P''}$ is perpendicular to $x = 2$ and $N$ is the intersection of $\overline{P'P''}$ with $x = 2$.

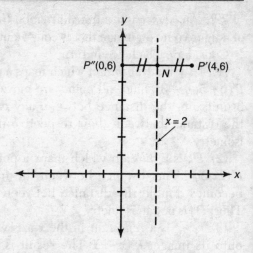

Therefore, a reflection in the line $x = 2$ replaces a point $P'(x,y)$ by its image $P''(x - 4,y)$. Therefore,

$$r_{x=2}P'(4,6) \rightarrow P''(4 - 4,6) \text{ or } P''(0,6)$$

Combining the results, $r_{x=2} \cdot r_{y\text{-axis}}P(-4,6) \rightarrow P''(0,6)$

The correct choice is (**4**).

**7.4.** The given function is $f(x) = \dfrac{4}{\sqrt{x+1}}$.

Over the set of real numbers, the domain of $f(x)$ is the subset of possible values that $x$ can take on and have the function defined.

The value of $x$ cannot be $-1$ since the denominator of $f(x)$ would then be 0, making $f(x)$ undefined. Also, if $x < -1$, the radicand of the denominator would be negative, thus making the denominator—and hence $f(x)$—an imaginary number. Therefore, $x$ must be greater than $-1$. The domain of $f(x)$ over the set of real numbers is $\{x | x > -1\}$.

The correct choice is (**4**).

**7.5.** The function is defined by: $\qquad f(x) = kx^2$

Since $f(2) = 12$, substitute 12 for $f(x)$ and 2 for $x$:
$$12 = k(2)^2$$
$$12 = 4k$$
$$3 = k$$

The correct choice is (**3**).

**7.6.** The given function is: $\qquad f(x) = \dfrac{x-4}{x+4}$

To find the value of $f(4a)$, substitute
$4a$ for $x$:
$$f(4a) = \dfrac{4a-4}{4a+4}$$

Factor the numerator by removing
4, the highest common factor of each
term. Factor the denominator by removing 4, the highest common factor
of each term:
$$= \dfrac{4(a-1)}{4(a+1)}$$

Divide out the common factor of 4
in the numerator and denominator:
$$= \dfrac{a-1}{a+1}$$

The correct choice is (**1**).

**7.7.** The given set is: $\qquad A = \{(1,2),(2,3)(3,4),(4,5)\}$

Interchanging the first and second
member of each ordered pair of set $A$
results in set $A^{-1}$, the inverse of $A$: $\qquad A^{-1} = \{(2,1),(3,2),(4,3),(5,4)\}$

A set of ordered pairs represents a function if no two ordered
pairs have the same first member but different second members.
Since set $A$ has this property and set $A^{-1}$ also has this property, $A$
and $A^{-1}$ are functions.

The correct choice is (**1**).

**7.8.** It is given that $f(x) = x - 3$ and $g(x) = x^2$. The notation $(f \circ g)(2)$ represents the composition of function $g$, evaluated for $x = 2$, followed by function $f$.

To evaluate $(f \circ g)(2)$ first find the value of $g(2)$ and then evaluate $f(x)$ using $g(2)$ as the value of $x$. Thus:

$$(f \circ g)(2) = f(g(2))$$
$$= f(4)$$
$$= 1$$

Since $g(x) = x^2$, $g(2) = 2^2 = 4$:
Since $f(x) = x - 3$, $f(4) = 4 - 3 = 1$:
The value of $(f \circ g)(2)$ is **1**.

**7.9.** It is given that $J = -2 + 5i$ and $K = 3 + 2i$.

  **a.** Using graph paper, draw coordinate axes in which the horizontal axis is the real axis and the vertical axis is the "imaginary" axis. To plot a complex number of the form $a + bi$, plot the ordered pair $(a,b)$. Plot and label $J(-2,5)$ and $K(3,2)$, as shown in the accompanying figure.

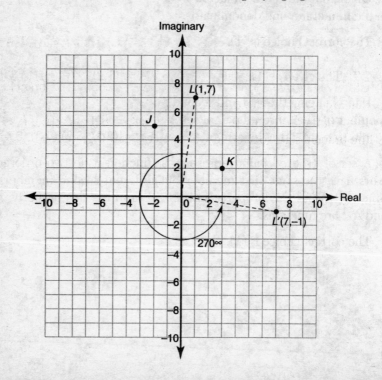

**b.** Since $L = J + K = (-2 + 5i) + (3 + 2i) = 1 + 7i$, plot $L(1,7)$, as shown in the accompanying figure.

**c.** In general, $R_{270°}(x,y) = (y,-x)$. Hence, after a counter-clockwise rotation of 270°, the image of $L(1,7)$ is $L'(7,-1)$, as shown in the accompanying figure.

**d.** For $L'(7,-1)$, $a = 7$ and $b = -1$, so $L'$, expressed as a complex number, is **$7 - i$**.

# 8. INVERSE VARIATION AND HYPERBOLAS

## 8.1 INVERSE VARIATION

- Variables $x$ and $y$ are inversely related if their product is constant.
- If $x$ and $y$ are inversely related, then $xy = k$, where $k \neq 0$.

## 8.2 RECTANGULAR (EQUILATERAL) HYPERBOLA

- The graph of an inverse variation $xy = k$ is a rectangular (or equilateral) hyperbola which consists of two disconnected branches that are asymptotic to the coordinate axes.
- If $k > 0$, the branches are located in Quadrants I and III; if $k < 0$, the branches are located in Quadrants II and IV.

# Practice Exercises

**8.1.** Given: $t$ varies inversely as $p$. If $p$ is divided by 2, then $t$ is

(1) increased by 2           (3) divided by 2

(2) decreased by 2        (4) multiplied by 2

**8.2.** *a* Draw and label the graph of the equation $xy = 6$ in the interval $-6 \leq x \leq 6$.

*b* On the same set of axes, draw and label the graph of the image of $xy = 6$ after a rotation of $90°$.

*c* Write the equation of the graph drawn in part *b*.

*d* On the same set of axes, draw and label the graph of the image of $xy = 6$ after a dilation of 2.

*e* Write the equation of the graph drawn in part *d*.

## Solutions

**8.1.** If $t$ varies inversely as $p$, then whenever the value of one of the two variables changes, the value of the other variable must change so that their product always remains the same. If $p$ is divided by 2, then $t$ must be *multiplied* by 2 so that their product does not change; that is, if $tp = k$ ($k > 0$), then

$$(2t)\left(\frac{p}{2}\right) = tp = k$$

The correct choice is (**4**).

**8.2. a.** The graph of an equation of the form $xy = k$ ($k \neq 0$) is a rectangular hyperbola. If $k > 0$, the hyperbola lies in Quadrants I and III since in these quadrants the product of $x$ and $y$ is positive.

On the closed interval $-6 \leq x \leq 6$ ($x \neq 0$), the graph of $xy = 6$ is in Quadrants I and III and has the $y$-axis as an asymptote, $x = 6$ as an endpoint in Quadrant I, and $x = -6$ as an endpoint in Quadrant III.

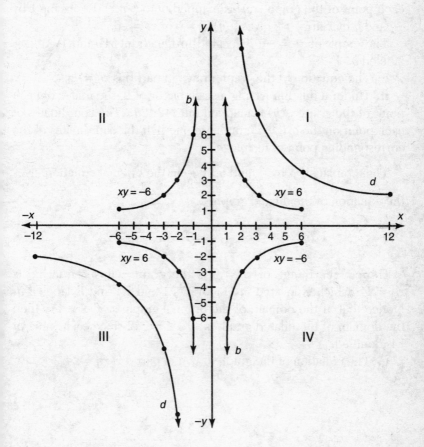

Locate some representative point on the graph by choosing integer values in the interval $-6 \leq x \leq 6$ (except 0) and finding the corresponding value of $y$ such that $xy = 6$. Organize these points in a table of values.

|   | Quadrant I |   |   |   | Quadrant III |   |   |   |
|---|---|---|---|---|---|---|---|---|
| $x$ | 1 | 2 | 3 | 6 | −1 | −2 | −3 | −6 |
| $y$ | 6 | 3 | 2 | 1 | −6 | −3 | −2 | −1 |

**b.** Under a counterclockwise rotation of 90° about the origin, each point of the graph $(x,y)$ is mapped onto $(-y,x)$. Replacing $x$ by $-y$ and $y$ by $x$ in $xy = 6$ gives $(-y)(x) = 6$ or $xy = -6$.

The graph of $xy = -6$ is located in Quadrants II and IV and is labeled $b$.

**c.** The equation of the graph drawn in part **b** is $xy = -6$.

**d.** Under a dilation having a scale factor of 2, the image of each point of the graph $(x,y)$ is mapped onto $(2x,2y)$. The coordinates of each point on the *original* graph are one-half the coordinates of the corresponding point of its image.

Thus, replacing $x$ by $\dfrac{x}{2}$ and $y$ by $\dfrac{y}{2}$ in the original equation gives the equation of the dilated graph:

$$xy = 6 \xrightarrow{\ D_2\ } \frac{x}{2} \cdot \frac{y}{2} = 6 \quad \text{or} \quad xy = 24$$

Hence, the image of $xy = 6$ is the graph whose equation is $xy = 24$, which is located in Quadrants I and III and is labeled $d$. Notice that, if the domain of the original graph is $-6 \leq x \leq 6$, then the domain of the dilated graph is $-12 \leq x \leq 12$ since each value of $x$ is doubled.

**e.** The equation of the graph drawn in part **d** is $xy = 24$.

# 9. EXPONENTIAL AND LOGARITHMIC FUNCTIONS

## 9.1 EXPONENTIAL FUNCTION

The equation $y = b^x$, where $b$ is different than 0 and 1, is called an exponential function. Its graph rises as $x$ increases when $b > 1$ and falls when $b < 1$.

## 9.2 LOGARITHMIC FUNCTION

The logarithmic function is the inverse of the exponential function. Hence, $y = \log_b x$ means $x = b^y$ provided $b$ is a positive number different than 1.

## 9.3 LOGARITHM LAWS

For all values of $x$ and $y$ for which these expressions are defined,

- $\log(xy) = \log x + \log y$
- $\log\left(\dfrac{x}{y}\right) = \log x - \log y$
- $\log(x^y) = y \log x$

## 9.4 SOLVING EXPONENTIAL EQUATIONS

- If both sides of an exponential equation can be expressed as a power of the same base, equate the exponents.
- If both sides of an exponential equation cannot be expressed as a power of the same base, then isolate the variable by taking the logarithm of each side of the equation. If $c^x = k$, then $x = \dfrac{\log k}{\log c}$, provided $c$ and $k$ are positive numbers.

# Practice Exercises

**9.1.** Solve for $x$: $27^{x+2} = 9^{2x-1}$

**9.2.** Which equation is equivalent to $y = 10^x$?

(1) $y = -10^{-x}$           (3) $y = \left(\dfrac{1}{10}\right)^{-x}$

(2) $y = 10^{-x}$            (4) $y = \left(\dfrac{1}{10}\right)^{x}$

**9.3.** *a* Sketch and label the graph of the equation $y = 2^x$.
    *b* Reflect the graph of the equation $y = 2^x$ in the $x$-axis. Label your answer $b$.
    *c* Reflect the graph of the equation $y = 2^x$ in the line $y = x$. Label your answer $c$.
    *d* Write an equation of the graph drawn in part $c$.

**9.4.** If $\log_x 5 = \dfrac{1}{2}$, find the value of $x$.

**9.5.** The expression $\dfrac{1}{2}\log a - 2\log b$ is equivalent to

(1) $\log \dfrac{\sqrt{a}}{b^2}$          (3) $\log \dfrac{a^2}{\sqrt{b}}$

(2) $\log \sqrt{ab}$           (4) $\log\left(\sqrt{a} - b^2\right)$

**9.6.** What is the $x$-intercept of the graph of the equation $y = \log_2 x$?

    (1) 1                         (3) 0

    (2) 2                         (4) 4

**9.7.** Given: $\log 3 = x$ and $\log 5 = y$.

    *a*  Express $\log \sqrt{\dfrac{3}{5}}$ in terms of $x$ and $y$.

    *b*  Express $\log 45$ in term of $x$ and $y$.

**9.8.** Using logarithms, find $w$ to the *nearest hundredth*.

$$5^{2w} + 9 = 40$$

## Solutions

**9.1.** The given equation, $27^{x+2} = 9^{2x-1}$, is an exponential equation that can be solved by expressing each side as a power of the same base and then equating the exponents.

| | |
|---|---|
| Given: | $27^{x+2} = 9^{2x-1}$ |
| Express each base as a power of 3: | $(3^3)^{x+2} = (3^2)^{2x-1}$ |
| Simplify: | $3^{3(x+2)} = 3^{2(2x-1)}$ |
| Equate the exponents: | $3(x + 2) = 2(2x - 1)$ |
| Remove the parentheses: | $3x + 6 = 4x - 2$ |

Collect like terms on the same side of the equation:  $8 = x$

The value of $x$ is **8**.

**9.2.** In general, if $b \neq 0$, then $b^x = \dfrac{1}{b^{-x}} = \left(\dfrac{1}{b}\right)^{-x}$. If $b = 10$, then

$$y = 10^x = \frac{1}{10^{-x}} = \left(\frac{1}{10}\right)^{-x}$$

The correct choice is **(3)**.

**9.3. a.** The equation $y = 2^x$ represents an exponential function. For $x = 0$, $y = 2^0 = 1$, so the graph crosses the $y$-axis at $(0,1)$.

For any value of $x$, $y$ is always positive, so the graph does not go below the $x$-axis. As $x$ grows larger in the positive direction, the graph slopes upward at an increasingly rapid rate. As $x$ becomes smaller in the negative direction, the graph approaches the $x$-axis as a horizonatal *asympote* but never quite reaches it.

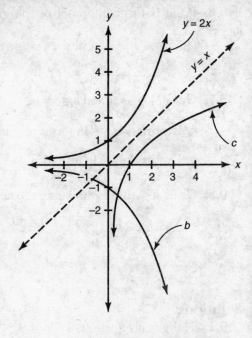

**b.** The reflection of the graph in the $x$-axis is labeled $b$. A reflection in the $x$-axis replaces each point $(x,y)$ of the graph with its image $(x,-y)$.

**c.** The line $y = x$ in $y = mx + b$ form is $y = 1x + 0$. Thus, its slope, $m$, is 1 and its $y$-intercept, $b$, is 0. The line passes through $(0,0)$ and is inclined $45°$ to the positive direction of the $x$-axis.

The reflection of $y = 2^x$ in the line $y = x$ is labeled $c$. A reflection in the line $y = x$ replaces each point $(x,y)$ of the graph with its image $(y,x)$.

**d.** As noted in part **c** above, a reflection in the line $y = x$ is equivalent to the transformation $(x,y)$; that is, $x$ is replaced by $y$ and $y$ is replaced by $x$. Therefore, an equation of the graph in part **c** is $x = 2^y$.

By definition, the logarithm of a number to a given base is the exponent to which that base must be raised to equal the number. In $x = 2^y$, the base, 2, is raised to the exponent, $y$, to equal the number, $x$. Therefore, $x = 2^y$ is equivalent to $y = \log_2 x$.

An equation of the graph in part **c** is $\boldsymbol{x = 2^y}$ or $\boldsymbol{y = \log_2 x}$.

**9.4.** Given: $\log_x 5 = \dfrac{1}{2}$

By definition, the logarithm of a number, $n$, to a base, $b$, is the exponent, $e$, to which the base must be raised to equal the number.

For $\log_x 5 = \dfrac{1}{2}$, the number $n = 5$, the base $b = x$,

and the exponent $e = \dfrac{1}{2}$: $\qquad x^{1/2} = 5$

$\qquad\qquad\qquad\qquad\qquad\qquad\qquad x = 25$

Square both sides of the equation: $\boldsymbol{x = 25}$.

**9.5.** The given expression is: $\qquad \dfrac{1}{2}\log a - 2\log b$

From the laws of logarithms, $\log \sqrt[n]{x} = \dfrac{1}{n}\log x$,

and $\log x^n = n\log x$: $\qquad\qquad\qquad \log a^{1/2} - \log b^2$

$\qquad\qquad\qquad\qquad\qquad\qquad\qquad \log \sqrt{a} - \log b^2$

From the laws of logarithms, $\log \dfrac{x}{y} = \log x - \log y$: $\qquad \log \dfrac{\sqrt{a}}{b^2}$

The correct choice is (**1**).

**9.6.** The $x$-intercept of a graph is the value of $x$ at the point where the graph crosses the $x$-axis, that is, where $y = 0$.

To find the $x$-intercept of $y = \log_2 x$, set $y = 0$: $\qquad 0 = \log_2 x$

By definition, the logarithm of a number to a given base is the exponent to which that base must be raised to equal the number. In $0 = \log_2 x$, the base is 2, the exponent (logarithm) is 0, and the number is $x$.

Therefore: $\qquad\qquad\qquad\qquad\qquad\qquad 2^0 = x$

But $2^0 = 1$: $\qquad\qquad\qquad\qquad\qquad\qquad 1 = x$

The $x$-intercept is 1.

The correct choice is (**1**).

**9.7.  a.** In general, $\log N^{1/r} = \frac{1}{r}\log N$ and $\log\left(\frac{a}{b}\right) = \log a - \log b$.

Use the power law of logarithms to obtain:

$$\log\sqrt{\frac{3}{5}} = \log\left(\frac{3}{5}\right)^{1/2} = \frac{1}{2}\log\left(\frac{3}{5}\right)$$

Next use the quotient law of logarithms:

$$= \frac{1}{2}(\log 3 - \log 5)$$

Replace $\log 3$ by $x$ and $\log 5$ by $y$ since this information is given:

$$= \frac{1}{2}(x - y)$$

Hence $\mathbf{\log\dfrac{3}{2} = \dfrac{1}{2}(x - y)}$.

**b.**  Express $\log 45$ in terms of the log of the product of powers of 3 and 5 and then simplify by using the power and product laws of logarithms. Since 45 can be factored as $3^2 \cdot 5$, rewrite $\log 45$ as follows:

$$\begin{aligned}\log 45 = \log(3^2 \cdot 5) &= \log 3^2 + \log 5 \\ &= 2\log 3 + \log 5 \\ &= 2x \quad + y\end{aligned}$$

$\mathbf{\log 45 = 2x + y}$

**9.8.**  The given equation, $5^{2w} + 9 = 40$, is an exponential equation that can be solved by isolating the variable term and then taking the logarithm of each side of the equation.

The given equation is:

$$5^{2w} + 9 = 40$$

Isolate the variable term:

$$5^{2w} = 31$$

Take the logarithm of each side of the equation:

$$\log(5^{2w}) = \log 31$$

Use the power law of logarithms:

$$2w\log 5 = \log 31$$

Solve for $w$:

$$w = \frac{\log 31}{2\log 5}$$

Use your scientific calculator to evaluate the logarithms:

$$= \frac{1.4914}{2(0.699)}$$

$$= 1.067$$

The value of $w$ to the *nearest hundredth* is **1.07**.

# 10. REGRESSION AND LINEAR CORRELATION

## 10.1 REGRESSION ANALYSIS

Fitting a line or curve to a set of data points is called **regression analysis**. A regression model can be classified according to the type of curve that is fitted to the data:

- A **linear regression** model has the form $y = ax + b$.
- An **exponential regression** model has the form $y = ab^x$.
- A **logarithmic regression** model has the form $y = a \ln x + b$.
- A **power regression** model has the form $y = ax^b$.

The regression feature of a graphing calculator allows you to choose the type of regression model and then calculates the constants $a$ and $b$ for the regression model selected. A regression model can be used to predict a $y$-value for a given $x$-value.

## 10.2 COEFFICIENT OF LINEAR CORRELATION

One type of regression model may fit data better than another type. The coefficient of linear correlation, denoted by $r$, is a number from $-1$ to $+1$ whose magnitude measures how closely the regression curve fits the data points and whose sign represents the direction of the relationship between the two variables.

- If $r > 0$, as one variable increases, the other variable increases. If $r < 0$, as one variable increases, the other variable decreases.
- If $|r| \approx 1$, the regression curve closely fits the data and there is a strong linear correlation between the two variables.
- If $|r| \approx 0$, the regression curve does not fit the data and there is no linear correlation between the two variables.

# Practice Exercises

**10.1.** The accompanying table shows the number of cricket chirps per minute recorded at various Fahrenheit temperatures.

    *a* Find a linear equation of the line that best fits the data. Estimate the regression coefficients to the *nearest hundredth*.

    *b* If Raymond counts 163 cricket chirps per minute, estimate the temperature to the *nearest tenth* of a Fahrenheit degree.

| Number of Cricket Chirps per Minute | Fahrenheit Temperature (degrees) |
|:---:|:---:|
| 50 | 52 |
| 78 | 56 |
| 90 | 61 |
| 115 | 72 |
| 190 | 85 |

**10.2.** The availability of leaded gasoline in New York State is decreasing, as shown in the accompanying table.

| Year | 1984 | 1988 | 1992 | 1996 | 2000 |
|---|---|---|---|---|---|
| Gallons Available (in thousands) | 150 | 124 | 104 | 76 | 50 |

*a* Determine a linear relationship for *x* (years) versus *y* (gallons available) based on the data given.

*b* If this relationship continues, during what year will leaded gasoline first become unavailable in New York State?

**10.3.** The amount A, in milligrams, of a dose of a drug remaining in the body is decreasing exponentially, as shown in the accompanying table.

| Elapsed Time (x hours) | 2 | 3 | 5 | 7 | 10 |
|---|---|---|---|---|---|
| Drug Dose Left in Body (y milligrams) | 6.3 | 5.1 | 3.4 | 2.1 | 1.2 |

*a* Based on the data given, determine an exponential function $y = ab^x$ that best fits the given data. Approximate *a* and *b* to the *nearest tenth*.

*b* If this relationship continues, find, to the *nearest tenth* of an hour, how long it will be before 0.5 milligram of the drug dose is left in the body.

## Solutions

**10.1. a.** Using your graphing calculator, enter the numbers of cricket chirps in list L1 and the corresponding Fahrenheit temperatures in list L2. Then select the linear regression option to get the display in the accompanying figure.

```
LinReg
 y=ax+b
 a=.2466912025
 b=39.39610022
 r²=.9623303708
 r=.9809843886
■
```

The equation of the line that best fits the data is $y = 0.25x + 39.4$.

**b.** If $x = 163$, then

$$y = 0.25x + 39.4 = 0.25(163) + 39.4 \approx \textbf{80.2 degrees}$$

**10.2.  a.** Using your graphing calculator,

- Enter the years in list L1 and the corresponding numbers of gallons available in list L2.
- Select option 4:LinReg(ax+b) from the STAT CALC menu.
- Assign Y1 to the linear regression equation by pressing [VARS] [▷] [ENTER] [ENTER]. This makes Y1 appear after LinReg(ax+b) in the display window.
- Press [ENTER] to produce the display in the accompanying figure.

```
LinReg
 y=ax+b
 a=-6.2
 b=12451.2
 r²=.9976641578
 r=-.9988313961
■
```

- Display the regression equation by pressing [Y=]:

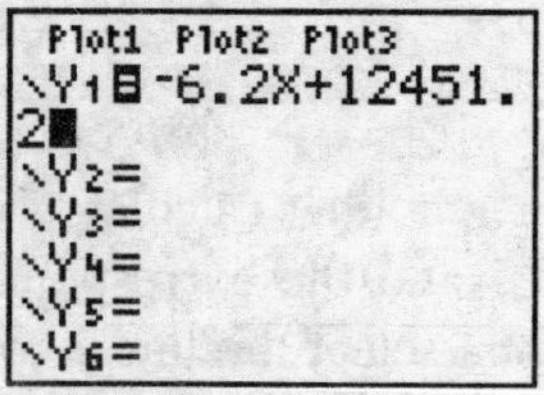

```
Plot1 Plot2 Plot3
\Y1 ▩ -6.2X+12451.
2■
\Y2=
\Y3=
\Y4=
\Y5=
\Y6=
```

Hence $y = -6.2x + 12451.2$.

      **b.** Create the table shown in the accompanying figure by setting the table start value at 2000 and then scrolling down until you reach a $y$-value less than 0:

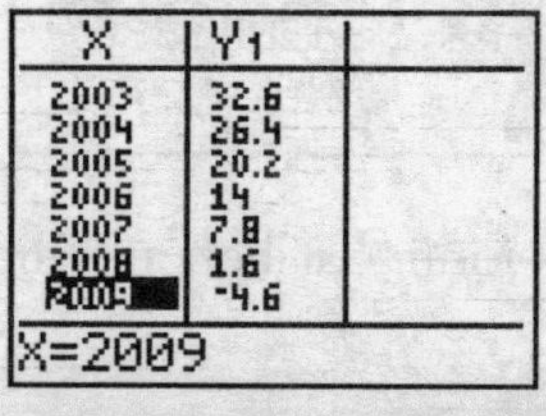

```
  X    │ Y1      │
2003   │ 32.6    │
2004   │ 26.4    │
2005   │ 20.2    │
2006   │ 14      │
2007   │ 7.8     │
2008   │ 1.6     │
2009   │ -4.6    │
X=2009
```

The $y$-value changes from positive to negative when the year changes from 2008 to 2009. This means that the year in which leaded gasoline will first become unavailable in New York State is **2008**.

**10.3. a.** Using your graphing calculator,

• Enter the elapsed times in list L1 and the corresponding numbers of milligrams of the drug dose in L2.

• Select option 0:ExpReg from the STAT CALC menu which produces the display in the accompanying figure.

```
ExpReg
 y=a*b^x
 a=9.534476689
 b=.8109822124
 r²=.9983662667
 r=-.9991827995
```

• The exponential function that best fits the given data is $y = 9.5(0.8)^x$.

    **b.** <u>Method 1: Use logarithms</u>

    Solve the exponential equation $y = 9.5(0.8)^x$ for $x$ when $y = 0.5 = \dfrac{1}{2}$:

$$\frac{1}{2} = 9.5(0.8)^x$$

$$\frac{1}{19} = (0.8)^x$$

$$\log\left(\frac{1}{19}\right) = \log\left[(0.8)^x\right]$$

$$-\log 19 = x \log 0.8$$

$$x = \frac{-\log 19}{\log 0.8} \approx 13.2 \text{ hours}$$

It takes **13.2 hours** for 0.5 milligram of the drug dose to be left in the body.

Method 2: Solve graphically

To solve the equation $0.5 = 9.5(0.8)^x$, graph $Y_1 = 0.5$ and $Y_2 = 9.5(0.8)^x$ in a friendly window such as $[0,18.8] \times [-1,1]$, as shown in the accompanying figure. Then use the intersect feature to find that at the point of intersection, $x \approx 13.2$.

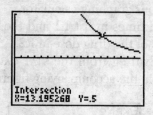

Intersection
X=13.195268  Y=.5

# 11.  SUMMATION NOTATION; STATISTICS

## 11.1  SUMMATION NOTATION

The symbol $\displaystyle\sum_{i=\text{starting value}}^{n=\text{end value}}$ (terms) is the Greek letter capital sigma

and represents the sum of the enclosed terms as the index variable $i$ successively takes on integer values in increments of 1 from its starting value to its end value.

## 11.2  STANDARD DEVIATION AND NORMAL CURVE

- The symbol $\sigma$ is the Greek letter lowercase sigma that represents the standard deviation of a set of data values. The standard deviation is a statistic that reflects how spread out the data are about the mean.

- When data are normally distributed, the data follow a bell-shaped curve. The area under a normal curve from $x_1$ to $x_2$ represents the percentage of scores that can be expected to fall within the interval from $x_1$ to $x_2$.

# Practice Exercises

**11.1.** Evaluate: $\displaystyle\sum_{k=1}^{4}\left(k^2+2\right)$

**11.2.** Evaluate: $\displaystyle 3\sum_{k=2}^{4}(k-2)^2$

**11.3.** In a normal distribution, $\bar{x}+2\sigma=80$ and $\bar{x}-2\sigma=40$ when $\bar{x}$ represents the mean and $\sigma$ represents the standard deviation. The standard deviation is

    (1) 10                 (3) 30

    (2) 20                 (4) 60

**11.4.** One thousand students took a test resulting in a normal distribution of the scores with a mean of 80 and a standard deviation of 5. Approximately how many students scored between 75 and 85?

    (1) 950              (3) 680

    (2) 815              (4) 475

**11.5.** The scores on a test approximate a normal distribution with a mean score of 72 and a standard deviation of 9. Approximately what percent of the students taking the test received a score greater than 90?

    (1) $2\frac{1}{2}\%$          (3) 10%

    (2) 5%             (4) 16%

## Solutions

**11.1.** $\sum_{k=1}^{4}\left(k^2+2\right)$ represents the sum of all values taken on by the expression $(k^2+2)$ as $k$ takes on successively the integral values from 1 to 4 inclusive:

$$\sum_{k=1}^{4}\left(k^2+2\right) = \left([1]^2+2\right)+\left([2]^2+2\right)+\left([3]^2+2\right)+\left([4]^2+2\right)$$
$$= (1+2)+(4+2)+(9+2)+(16+2)$$
$$= 38$$

The value is **38**.

**11.2.** $\sum_{k=2}^{4}\left(k-2\right)^2$ represents the summation of all terms of the form $(k-2)^2$ as $k$ takes on successively the integral values from 2 to 4 inclusive:

**11.3.** The symbol $\bar{x}$ represents the mean of the normal distribution, and $\sigma$ represents the standard deviation. It is given that:

$$\bar{x}+2\sigma = 80$$
$$\bar{x}-2\sigma = 40$$

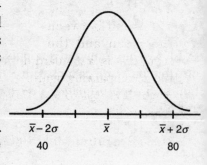

To eliminate $\bar{x}$, subtract the second equation from the first:

$$\bar{x}+2\sigma = 80$$
$$-\bar{x}+2\sigma = -40$$
$$\overline{\quad 4\sigma = 40 \quad}$$
$$\sigma = 10$$

The correct choice is **(1)**.

**11.4.** If the mean is 80 and the standard deviation is 5, the scores between 75 and 85 represent the range within 1 standard deviation of the mean. In a normal distribution, 68% of the scores fall within 1 standard deviation of the mean. Since 1000 students took the test, 68% of 1000, or 680 students, should score between 75 and 85.

The correct choice is **(3)**.

**11.5.** It is given that the scores on a test approximate a normal distribution with a mean score of 72 and a standard deviation of 9.

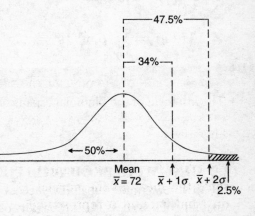

- Since $90 = 72 + 18 = 72 + (2 \cdot 9)$, a score of 90 is 2 standard deviations above the mean.
- When data are normally distributed, 50% of the scores are below the mean and about 47.5% of the scores fall between the mean and the score that is 2 standard deviations above the mean, as shown in the accompanying figure.
- About $50\% + 47.5\% = 97.5\%$ of the test scores are less than or equal to 90.

- Hence, approximately $100\% - 97.5\% = 2.5\%$ or $2\frac{1}{2}\%$ of the scores are greater than 90.

The correct choice is **(1)**.

# 12. BINOMIAL THEOREM AND PROBABILITY

## 12.1 COMBINATIONS AND THE BINOMIAL THEOREM

- The number of combinations of $n$ items taken $r$ at a time is $_nC_r$, where

$$_nC_r = \frac{n!}{(n-r)! \cdot r!}$$

- If $n$ is a positive integer, then

$$(x+y)^n = \sum_{r=0}^{n} {_nC_r}\, x^{n-r} y^r = {_nC_0}\, x^n y^0 + {_nC_1}\, x^{n-1} y^1$$
$$+ {_nC_2}\, x^{n-2} y^2 + \cdots + {_nC_n}\, x^0 y^n$$

- The $k$th term of the binomial expansion of $(x+y)^n$ is

$$_nC_{k-1}\, x^{n-(k-1)} y^{k-1}$$

## 12.2 BINOMIAL PROBABILITIES

In a two-outcome probability success in which $p$ represents the probability of a success, the probability of obtaining $r$ successes in $n$ trials $(r \le n)$ is given by the expression

$$_nC_r\, p^r (1-p)^{n-r}$$

# Practice Exercises

**12.1.** Find, in simplest form, the middle term in the expansion of $\left(x^2 + \dfrac{1}{x}\right)^6$.

**12.2.** Assume that in the United States $\dfrac{1}{5}$ of all cars are red. Suppose you are driving down a highway and you pass 6 cars.

    *a* What is the probability that *at most* one of the cars you pass is red?

    *b* What is the probability that *at least* four of the cars you pass are red?

## Solutions

**12.1.** In general, the expansion of a binomial of the form $(a + b)^n$ has $n + 1$ terms of which the $k$th term is $_nC_{k-1}a^{n-(k-1)}b^{k-1}$.

Since the expansion of the given expression, $\left(x^2 + \dfrac{1}{x}\right)^6$, has $6 + 1$ = 7 terms, the middle term in the expansion of the binomial is the fourth term.

To find the fourth term of the expansion of $\left(x^2 + \dfrac{1}{x}\right)^6$, use the expression $_nC_{k-1}a^{n-(k-1)}b^{k-1}$, where $n = 6$, $k = 4$, $a = x^2$, and $b = \dfrac{1}{x}$:

$$_nC_{k-1}a^{n-(k-1)}b^{k-1} = {}_6C_{4-1}\left(x^2\right)^{6-(4-1)}\left(\frac{1}{x}\right)^{4-1}$$

$$= {}_6C_3\left(x^2\right)^3\left(\frac{1}{x}\right)^3$$

Use a scientific calculator to
obtain $_6C_3 = 20$:

$$= 20\left(x^6\right)\left(\frac{1}{x^3}\right)$$

$$= 20x^3$$

The middle term in the expansion of $\left(x^2 + \dfrac{1}{x}\right)^6$ is **$20x^3$**.

**12.2.** You are asked to assume that in the United States $\dfrac{1}{5}$ of all cars are red and that, when driving down a highway, you pass 6 cars.

Finding the probability that $k$ of the $n$ cars passed are red may be considered a two-outcome probability experiment in which

$$p(k \text{ of } n \text{ cars passed are red}) = {}_nC_r p^r q^{n-r}$$

where $p$ = probability of passing a red car = $\dfrac{1}{5}$ and $q$ = probability of not passing a red car = $1 - \dfrac{1}{5} = \dfrac{4}{5}$.

**a.** To find the probability that *at most* one of the cars you pass is red, calculate the sum of the probabilities that you will pass 0 cars and 1 car.

$$P(0 \text{ red}) = {}_6C_0\left(\frac{1}{5}\right)^0\left(\frac{4}{5}\right)^{6-0} = \quad 1 \cdot 1 \cdot \left(\frac{4}{5}\right)^6 = \frac{4096}{15625}$$

$$+ P(1 \text{ red}) = {}_6C_1\left(\frac{1}{5}\right)^1\left(\frac{4}{5}\right)^{6-1} = 6 \cdot \left(\frac{1}{5}\right) \cdot \left(\frac{1024}{3125}\right) = \frac{6144}{15625}$$

$$P(0 \text{ red}) + P(1 \text{ red}) = \frac{10240}{15625}$$

The probability that *at most* one of the cars you pass is red is $\dfrac{10240}{15625}$.

    **b.** To find the probability that *at least* four of the cars you pass is red, calculate the sum of the probabilities of passing 4, 5, and 6 cars.

$$P(4 \text{ red}) = {}_6C_4\left(\frac{1}{5}\right)^4\left(\frac{4}{5}\right)^{6-4} = 15 \cdot \frac{1}{625} \cdot \frac{16}{25} = \frac{240}{15625}$$

$$+\, P(5 \text{ red}) = {}_6C_5\left(\frac{1}{5}\right)^5\left(\frac{4}{5}\right)^{6-5} = 6 \cdot \frac{1}{3125} \cdot \frac{4}{5} = \frac{24}{15625}$$

$$+\, P(6 \text{ red}) = {}_6C_6\left(\frac{1}{5}\right)^6\left(\frac{4}{5}\right)^{6-6} = 1 \cdot \frac{1}{15625} \cdot 1 = \frac{1}{15625}$$

$$P(4 \text{ red}) + P(5 \text{ red}) + P(6 \text{ red}) = \frac{265}{15625}$$

The probability that *at least* four of the cars you pass are red is $\dfrac{265}{15625}$.

# 13. GENERAL ANGLE; RADIAN MEASURE; ARC LENGTH; AREA

## 13.1 RADIAN MEASURE, ARC LENGTH, AND AREA

- To change from degree measure to radian measure, multiply the number of degrees by $\dfrac{\pi}{180°}$. To change from radian measure to degree measure, multiply the number of radians by $\dfrac{180°}{\pi}$.

- The length of an arc intercepted by a central angle of $\theta$ radians is radius $\times \theta$.

- The area of a sector of a circle bounded by two radii and the arc opposite a central angle of $n°$ is $\dfrac{n}{360} \times \pi \times (\text{radius})^2$.

## 13.2 TRIGONOMETRIC FUNCTIONS OF 30°, 45°, AND 60°

The exact values of the basic trigonometric functions of 30°, 45°, and 60° are summarized in the accompanying table. To find the corresponding values of the cosecant, secant, and cotangent, use the reciprocal identities:

$$\csc x = \frac{1}{\sin x}, \sec x = \frac{1}{\cos x}, \text{ and } \cot x = \frac{1}{\tan x}$$

| $x$ | $\sin x$ | $\cos x$ | $\tan x$ |
|---|---|---|---|
| 30° | $\dfrac{1}{2}$ | $\dfrac{\sqrt{3}}{2}$ | $\dfrac{1}{\sqrt{3}}$ |
| 45° | $\dfrac{\sqrt{2}}{2}$ | $\dfrac{\sqrt{2}}{2}$ | $1$ |
| 60° | $\dfrac{\sqrt{3}}{2}$ | $\dfrac{1}{2}$ | $\sqrt{3}$ |

## 13.3  GENERAL ANGLE

- If an angle $\theta$ is in standard position and $P(x, y)$ is any point on the terminal side of angle $\theta$, where $r = \sqrt{x^2 + y^2}$, then for all values of $\theta$ for which the trigonometric functions are defined:

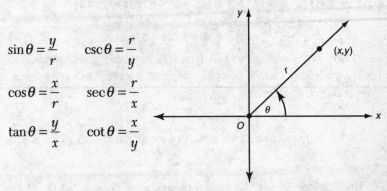

$$\sin\theta = \frac{y}{r} \qquad \csc\theta = \frac{r}{y}$$

$$\cos\theta = \frac{x}{r} \qquad \sec\theta = \frac{r}{x}$$

$$\tan\theta = \frac{y}{x} \qquad \cot\theta = \frac{x}{y}$$

- The algebraic signs of the trigonometric functions depend on the signs of $x$ and $y$ in the particular quadrant in which the terminal side of angle $\theta$ is located.
- The accompanying table shows how to express the trigonometric function of an angle as a function of an *acute* angle, where $x_{ref}$ is the acute angle formed by the terminal side of the general angle $x$ and the $x$-axis. For example:

$$\sin 210° = -\sin 30°, \cos 135° = -\cos 45°, \text{ and } \tan 250° = \tan 70°$$

|  | Quadrant I: all functions are + | Quadrant II: sin x is + | Quadrant III: tan x is + | Quadrant IV: cos x is + |
|---|---|---|---|---|
| $\sin x$ (and $\csc x$) | $\sin x = \sin x_{ref}$ | $\sin x = \sin x_{ref}$ | $\sin \overline{x} = -\sin x_{ref}$ | $\sin x = -\sin x_{ref}$ |
| $\cos x$ (and $\sec x$) | $\cos x = \cos x_{ref}$ | $\cos x = -\cos x_{ref}$ | $\cos x = -\cos x_{ref}$ | $\cos x = \cos x_{ref}$ |
| $\tan x$ (and $\cot x$) | $\tan x = \tan x_{ref}$ | $\tan x = -\tan x_{ref}$ | $\tan x = \tan x_{ref}$ | $\tan x = -\tan x_{ref}$ |

# Practice Exercises

**13.1.** Express in degree measure an angle of $\dfrac{2\pi}{5}$ radians.

**13.2.** Express 140° in radian measure.

**13.3.** In a circle of radius 8, find the length of the arc intercepted by a central angle of 1.5 radians.

**13.4.** In the accompanying diagram of a unit circle, the ordered pair $(x,y)$ represents the point where the terminal side of $\theta$ intersects the unit circle.

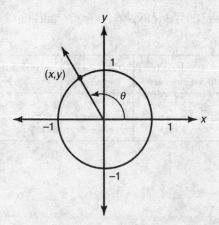

If $\theta = \dfrac{3\pi}{4}$, what is the value of $x$?

(1)  1

(3)  $-\dfrac{\sqrt{2}}{2}$

(2)  $-\dfrac{1}{2}$

(4)  $\dfrac{\sqrt{3}}{2}$

**13.5.**  In the accompanying diagram of a unit circle, $\overline{BA}$ is tangent to circle $O$ at $A$, $\overline{CD}$ is perpendicular to the $x$-axis, and $OC$ is a radius.

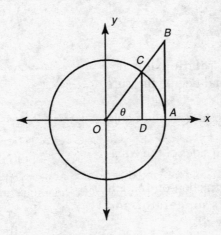

Which distance represents $\sin \theta$?
(1)  $OD$                     (3)  $BA$
(2)  $CD$                     (4)  $OB$

**13.6.**  If $\sin \theta = \cos \theta$, in which quadrants can angle $\theta$ terminate?
      (1)  I, II                     (3)  I, III
      (2)  II, III                   (4)  I, IV

**13.7.**  Which expression is equivalent to $\sin(-120°)$?
    (1)  $\sin 60°$                (3)  $\cos 30°$
    (2)  $-\sin 60°$               (4)  $-\sin 30°$

## Solutions

**13.1.** The relationship between radian and degree measure is:

$$\pi \text{ radians} = 180°$$

Multiply both sides of this equation by $\frac{2}{5}$:

$$\frac{2\pi}{5} \text{ radians} = \frac{2}{5}(180°)$$

$$\frac{2\pi}{5} \text{ radians} = \frac{360°}{5}$$

$$\frac{2\pi}{5} \text{ radians} = 72°$$

$$\frac{2\pi}{5} \textbf{ radians} = \textbf{72°}.$$

**13.2.** The relationship between radian and degree measure is:

$$\pi \text{ radians} = 180°$$

Divide both sides of the equation by 180:

$$\frac{\pi}{180} \text{ radians} = 1°$$

Multiply both sides of the equation by 140:

$$\frac{140\pi}{180} \text{ radians} = 140°$$

Reduce the fraction on the left by dividing its numerator and denominator by 20:

$$\frac{7\pi}{9} \text{ radians} = 140°$$

$$\textbf{140°} = \frac{7\pi}{9} \textbf{ radians}.$$

**13.3.** The length of an arc equals the radius of the circle multiplied by the measure of its central angle in radians. Here the length of the arc = 8(1.5) = 12.0.

The length of the arc is **12**.

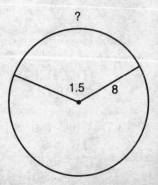

**13.4.** It is given that, in the accompanying diagram of a unit circle, the ordered pair $(x,y)$ represents the point where the terminal side of $\theta$ intersects the unit circle.

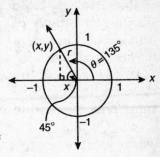

If $\theta = \dfrac{3\pi}{4} = 135°$, use the cosine ratio to find the value of $x$ letting $\cos 135° = -\cos 45° = -\dfrac{\sqrt{2}}{2}$ and $r = 1$:

$$\cos\theta = \frac{x}{r}$$

$$-\cos 45° = \frac{x}{1}$$

$$-\frac{\sqrt{2}}{2} = x$$

The correct choice is **(3)**.

**13.5.** It is given that, in the accompanying diagram of a unit circle, $\overline{BA}$ is tangent to circle $O$ at $A$, $\overline{CD}$ is perpendicular to the $x$-axis, and $\overline{OC}$ is a radius.

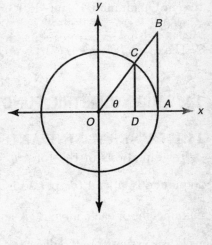

Since circle $O$ is a unit circle, the length of radius $\overline{OC}$ is 1. In right triangle $ODC$:

$$\sin\theta = \frac{\text{side opposite } \angle\theta}{\text{hypotenuse}}$$

$$= \frac{CD}{OC}$$

$$= \frac{CD}{1}$$

$$= CD$$

The correct choice is **(2)**.

**13.6.** Make a table showing the signs of $\sin\theta$ and $\cos\theta$ in each of the four quadrants:

If $\sin\theta = \cos\theta$, the two must have the same sign. This is possible

|  | Quadrant | | | |
|---|:---:|:---:|:---:|:---:|
|  | **I** | **II** | **III** | **IV** |
| $\sin\theta$ | + | + | − | − |
| $\cos\theta$ | + | − | − | + |

only in Quadrant I (where both are positive) or in Quadrant III (where both are negative).

The correct choice is **(3)**.

**13.7.** Represent −120° in standard position. Since −120° is negative, its terminal side is located by rotating 120° in a *clockwise* direction from the initial position. Thus, the terminal side will lie in Quadrant III, and the reference angle will be 60°.

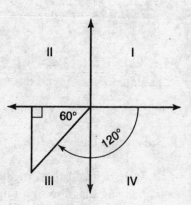

Since the sine function is negative in Quadrant III, $\sin(-120°) = -\sin 60°$.

The correct choice is **(2)**.

# 14. TRIGONOMETRIC FUNCTIONS AND GRAPHS

## 14.1 GENERAL SINE AND COSINE CURVES

For equations of the form $y = a\sin bx$ and $y = a\cos bx$, the amplitude is $|a|$ and the period is $\left|\dfrac{2\pi}{b}\right|$. This means that the maximum height of each curve is $|a|$, and each graph completes one full cycle as $x$ varies from 0 radians to $\dfrac{2\pi}{b}$ radians.

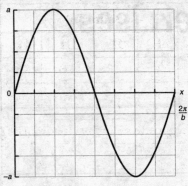

Graph of $y = a \sin bx$, where $a, b > 0$.

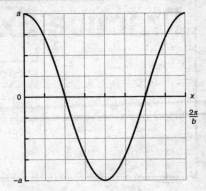

Graph of $y = a \cos bx$, where $a, b > 0$.

## 14.2 BASIC TANGENT CURVE

The graph of $y = \tan x$ has no amplitude with vertical asymptotes at odd multiples of $\dfrac{\pi}{2}$ radians.

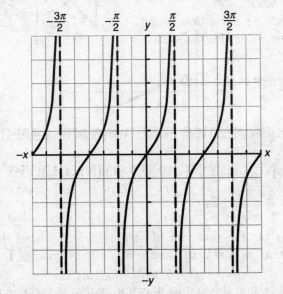

Graph of $y = \tan x$ with vertical asymptotes at $\pm\dfrac{\pi}{2}$ and $\pm\dfrac{3\pi}{2}$.

# Practice Exercises

**14.1.** What is the amplitude of the graph of the equation $y = 2\cos 3x$?

(1) $\dfrac{2\pi}{3}$             (3) 3

(2) 2                  (4) $6\pi$

**14.2.** Which is an equation of the graph shown below?

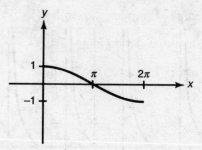

(1) $y = \cos\dfrac{1}{2}x$        (3) $y = \sin\dfrac{1}{2}x$

(2) $y = \cos 2x$           (4) $y = \sin 2x$

**14.3.** *a* On the same set of axes, sketch and label the graphs of the equations $y = 2\sin x$ and $y = \cos 2x$ as $x$ varies from $-\pi$ to $\pi$ radians.

       *b* Using the graphs drawn in part *a*, determine the value of $x$ in the interval $-\pi \le x \le \pi$ such that $2\sin x - \cos 2x = 3$.

## Solutions

**14.1.** If an equation of a cosine curve is in the form $y = a \cos bx$, then $a$ represents the amplitude (maximum height of the graph) and $\dfrac{360°}{b}$ represents the period.

The given equation, $y = 2 \cos 3x$, is in the form $y = a \cos bx$, with $a = 2$ and $b = 3$. Therefore, the amplitude of the graph is 2.

The correct choice is **(2)**.

**14.2.** If the equation on the graph were of the form $y = a \sin bx$, $(0,0)$ would be a point on the graph since $\sin 0° = 0$. Since $(0,0)$ is not on the graph, choice (3), $y = \sin \dfrac{1}{2}x$, and choice (4),

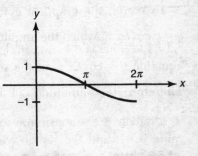

$y = \sin 2x$, can be ruled out.

Choices (1) and (2) are both of the form $y = a \cos bx$. If the graph of an equation is in the form $y = a \cos bx$, its amplitude or maximum height is $a$, and its period (number of radians in one complete cycle) is $\dfrac{2\pi}{b}$. The amplitude shown on the graph is 1; hence $a = 1$. There is only one-half of a complete cycle from $x = 0$ to $x = 2\pi$; hence the period is twice this range, or $4\pi$.

$$\frac{2\pi}{b} = 4\pi, \text{ or } 2\pi = 4\pi b, \text{ or } b = \frac{1}{2}$$

Thus, the equation is $y = \cos \dfrac{1}{2}x$.

The correct choice is **(1)**.

**14.3. a.** To sketch the graph of $y = 2 \sin x$, first determine the *amplitude* and *period* and then use these values in making the sketch. The amplitude is the height or maximum value of the curve. The period is the number of degrees (or radians) through which one complete cycle of the curve will extend.

In the standard equation $y = a \sin bx$, the amplitude is $a$ and the period is $\frac{360°}{b}$. The equation $y = 2 \sin x$ is in the standard form *with* $a = 2$ and $b = 1$. Therefore, the amplitude is 2 and the period is $\frac{360°}{1}$ or 360°. The curve has a maximum value of 2 and a minimum value of −2, and one complete cycle extends 360° (or $2\pi$ radians). Since the question asks for values of $x$ from $-\pi$ to $\pi$ radians, the graph will cover exactly one cycle.

To sketch the graph of $y = \cos 2x$, use the standard equation $y = a \cos bx$ in which the amplitude is $a$ and the period is $\frac{360°}{b}$. The equation $y = \cos 2x$ is in the standard form with $a = 1$ and $b = 2$. Therefore, the amplitude is 1 and the period is $\frac{360°}{2}$ or 180° (that is, $\pi$ radians). The maximum value of the curve is 1, the minimum value is −1, and the sketch will show two complete cycles in the interval from $x = -\pi$ to $x = \pi$.

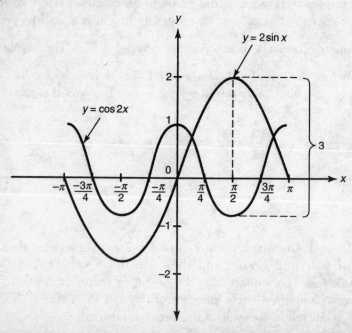

**b.** In order for $2\sin x - \cos 2x$ to equal 3, the ordinate ($y$-value) on the graph of $y = 2\sin x$ must exceed the ordinate on the graph of $y = \cos 2x$ by 3. This occurs when $x = \dfrac{\pi}{2}$.

$$x = \frac{\pi}{2}.$$

# 15. TRIGONOMETRIC IDENTITIES AND EQUATIONS

## 15.1 BASIC TRIGONOMETRIC IDENTITIES

Sometimes it is necessary to change from one trigonometric function to another by using a Pythagorean, reciprocal, or quotient identity:

- Pythagorean identities:

  $\sin^2 A + \cos^2 A = 1$
  $\tan^2 A + 1 = \sec^2 A$
  $\cot^2 A + 1 = \csc^2 A$

- Reciprocal identities:

  $\sin A = \dfrac{1}{\csc A}$ and $\csc A = \dfrac{1}{\sin A}$

  $\cos A = \dfrac{1}{\sec A}$ and $\sec A = \dfrac{1}{\cos A}$

  $\tan A = \dfrac{1}{\cot A}$ and $\cot A = \dfrac{1}{\tan A}$

- Quotient identities:

  $\tan A = \dfrac{\sin A}{\cos A}$ and $\cot A = \dfrac{\cos A}{\sin A}$

## 15.2 IDENTITIES FOR SUM, DIFFERENCE, DOUBLE ANGLE, AND HALF-ANGLE IDENTITIES

See formulas on pages 30 to 35.

# Practice Exercises

**15.1.** What is the value of $x$ in the interval $90° \leq x \leq 180°$ that satisfies the equation $\sin x + \sin^2 x = 0$?

      (1) $90°$                  (3) $180°$

      (2) $135°$             (4) $270°$

**15.2.** For all values of $\theta$ for which the expression is defined, $\dfrac{\sec \theta}{\cos \theta}$ is equivalent to

      (1) $\sin \theta$             (3) $\tan \theta$

      (2) $\cos \theta$            (4) $\cot \theta$

**15.3.** The expression $\cos y(\csc y - \sec y)$ is equivalent to

      (1) $\cot y - 1$         (3) $1 - \tan y$

      (2) $\tan y - 1$         (4) $-\cos y$

**15.4.** If $\sin A = \dfrac{3}{5}$, find $\cos 2A$.

**15.5.** $\cos 70° \cos 40° - \sin 70° \sin 40°$ is equivalent to

      (1) $\cos 30°$          (3) $\cos 110°$

      (2) $\cos 70°$          (4) $\sin 70°$

**15.6.** If $\sin\alpha = \dfrac{4}{5}$, $\tan\beta = \dfrac{5}{12}$, and $\alpha$ and $\beta$ are first-quadrant angles, what is the value of $\sin(\alpha+\beta)$?

(1) $\dfrac{63}{65}$  (3) $\dfrac{33}{65}$

(2) $-\dfrac{33}{65}$  (4) $-\dfrac{63}{65}$

**15.7.** If $\cos A = \dfrac{1}{3}$, then the positive value of $\tan\dfrac{1}{2}A$ is

(1) $\sqrt{2}$  (3) $\dfrac{\sqrt{3}}{3}$

(2) $\sqrt{3}$  (4) $\dfrac{\sqrt{2}}{2}$

**15.8.** Find, to the *nearest degree*, all values of $\theta$ in the interval $0° \le \theta < 180°$ that satisfy the equation $3\tan^2\theta + \dfrac{1}{\cot\theta} = 2$.

## Solutions

**15.1.** The given equation is a trigonometric equation:

$$\sin x + \sin^2 x = 0$$

Factor out the common factor, $\sin x$, on the left side:

$$\sin x(1 + \sin x) = 0$$

If the product of two factors is 0, either factor can equal 0:

$$\sin x = 0 \text{ or } 1 + \sin x = 0$$
$$x = 0°, \ 180° \text{ or } \sin x = -1$$
$$x = 270°$$

The only value of $x$ in the interval $90° \le x \le 180°$ is $180°$.

The correct choice is **(3)**.

**15.2.** The given expression is:

$$\dfrac{\sec\theta}{\csc\theta}$$

Since $\sec\theta = \dfrac{1}{\cos\theta}$ and $\csc\theta = \dfrac{1}{\sin\theta}$:

$$\dfrac{\dfrac{1}{\cos\theta}}{\dfrac{1}{\sin\theta}}$$

By definition,

$$\dfrac{\dfrac{1}{\cos\theta}}{\dfrac{1}{\sin\theta}} = \dfrac{1}{\cos\theta} \div \dfrac{1}{\sin\theta} = \dfrac{1}{\cos\theta} \times \dfrac{\sin\theta}{1} = \dfrac{\sin\theta}{\cos\theta}:$$

$$\dfrac{\sin\theta}{\cos\theta}$$

From one of the quotient identities, $\dfrac{\sin\theta}{\cos\theta} = \tan\theta$:          $\tan\theta$

The correct choice is **(3)**.

**15.3.** The given expression is:          $\cos y(\csc y - \sec y)$

Since $\csc y = \dfrac{1}{\sin y}$ and $\sec y = \dfrac{1}{\cos y}$:          $\cos y\left(\dfrac{1}{\sin y} - \dfrac{1}{\cos y}\right)$

Remove the parentheses by applying the Distributive Law:          $\dfrac{\cos y}{\sin y} - 1$

But $\dfrac{\cos y}{\sin y} = \cot y$:          $\cot y - 1$

The correct choice is **(1)**.

**15.4.** From the Functions of the Double Angle:          $\cos 2A = 1 - 2\sin^2 A$

Given $\sin A = \dfrac{3}{5}$:          $\cos 2A = 1 - 2\left(\dfrac{3}{5}\right)^2$

$$\cos 2A = 1 - 2\left(\dfrac{9}{25}\right)$$

$$\cos 2A = 1 - \dfrac{18}{25}$$

$$\cos 2A = \dfrac{25}{25} - \dfrac{18}{25} = \dfrac{7}{25}$$

$$\cos 2A = \frac{7}{25}.$$

**15.5.** The given expression is:         $\cos 70° \cos 40° - \sin 70° \sin 40°$

From the Functions of the Sum of Two Angles:     $\cos(A + B) = \cos A \cos B - \sin A \sin B$

Let $A = 70°$, and $B = 40°$:       $\cos(70° + 40°) = \cos 70° \cos 40° - \sin 70° \sin 40°$

$$\cos(110°) = \cos 70° \cos 40° - \sin 70° \sin 40°$$

The correct choice is **(3)**.

**15.6.** Given: $\sin \alpha = \dfrac{4}{5}$, $\tan \beta = \dfrac{5}{12}$, and $\alpha$ and $\beta$ are first-quadrant angles.

Since $\alpha$ is a first-quadrant angle, it can be represented as an acute angle in a right triangle whose opposite leg is 4 and whose hypotenuse is 5. From the 3-4-5 right-triangle relationship, the adjacent leg is 3.

Since $\beta$ is a first-quadrant angle, it can be represented as an acute angle in a right triangle whose opposite leg is 5 and whose adjacent leg is 12. From the 5-12-13 right-triangle relationship, the hypotenuse is 13.

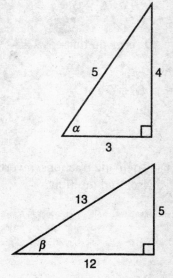

From the Functions of the Sum of Two Angles:      $\sin(\alpha + \beta) = \sin \alpha \cos \beta + \cos \alpha \sin \beta$

$$\cos\alpha = \frac{\text{adjacent leg}}{\text{hypotenuse}} = \frac{3}{5} \quad \text{and} \quad \sin\beta = \frac{\text{opposite leg}}{\text{hypotenuse}} = \frac{5}{13}$$

$$\text{and} \quad \cos\beta = \frac{\text{adjacent leg}}{\text{hypotenuse}} = \frac{12}{13}$$

Therefore:
$$\sin(\alpha+\beta) = \left(\frac{4}{5}\right)\left(\frac{12}{13}\right) + \left(\frac{3}{5}\right)\left(\frac{5}{13}\right)$$

$$= \frac{48}{65} + \frac{15}{65}$$

$$= \frac{63}{65}$$

The correct choice is **(1)**.

**15.7.** From the Functions of the Half-Angle:

$$\tan\frac{1}{2}A = \pm\sqrt{\frac{1-\cos A}{1+\cos A}}$$

It is given that $\cos A = \frac{1}{3}$:

$$\tan\frac{1}{2}A = \pm\sqrt{\frac{1-\frac{1}{3}}{1+\frac{1}{3}}}$$

$$= \pm\sqrt{\frac{\frac{2}{3}}{\frac{4}{3}}}$$

To divide fractions, invert the divisor and multiply:

$$\tan\frac{1}{2}A = \pm\sqrt{\frac{2}{3} \times \frac{3}{4}}$$

$$= \pm\sqrt{\frac{2}{4}}$$

**15.8.** The given equation is:

$$3\tan^2\theta + \frac{1}{\cot\theta} = 2$$

Since $\frac{1}{\cot\theta} = \tan\theta$:

$$3\tan^2\theta + \tan\theta = 2$$

This is a *quadratic equation*; rearrange it so that all terms are on one side equal to 0:

$$3\tan^2\theta + \tan\theta - 2 = 0$$

The left side is a *quadratic trinomial* that can be factored into the product of two binomials; be sure to check that the product of the inner terms of the binomials added to the product of the outer terms equals the middle term, $+\tan\theta$, of the original trinomial:

$$-2\tan\theta = \text{inner product}$$
$$(3\tan\theta - 2)(\tan\theta + 1) = 0$$
$$+3\tan\theta = \text{outer product}$$

Since $(+3\tan\theta) + (-2\tan\theta) = +\tan\theta$, these are the correct factors:

$$(3\tan\theta - 2)(\tan\theta + 1) = 0$$

If the product of two factors is 0, either factor can equal 0:

$$3\tan\theta - 2 = 0 \quad \text{or} \quad \tan\theta + 1 = 0$$
$$3\tan\theta = 2 \qquad\qquad \tan\theta = -1$$
$$\tan\theta = \frac{2}{3}$$

The question calls for values of $\theta$ in the interval $0° \leq \theta < 180°$, that is, in Quadrant I or II. For $\tan\theta = -1$, $\theta$ must be in Quadrant II since $\tan\theta$ is negative; $\tan 45° = 1$, and therefore $\theta = 135°$. For $\tan\theta = \frac{2}{3}$ or 0.6667, a scientific calculator shows that $\tan 33°40' = 0.6661$; to the *nearest degree*, $\theta = 34°$.

$$\theta = 34°, 135°$$

# 16.  TRIGONOMETRIC LAWS AND APPLICATIONS

## 16.1  AREA OF A TRIANGLE

The area of a triangle can be expressed in terms of the measures of two sides and the sine of their included angle. The area, $K$, of $\triangle ABC$ is

$$K = \frac{1}{2}ab\sin C$$

## 16.2  LAW OF SINES

The Law of Sines relates the sines of any two angles of a triangle to the lengths of the sides opposite these angles. In $\triangle ABC$:

$$\frac{a}{\sin A} = \frac{b}{\sin B} = \frac{c}{\sin C}$$

## 16.3  LAW OF COSINES

The Law of Cosines relates the lengths of the three sides of any triangle to the cosine of one of its angles. In $\triangle ABC$:

$$a^2 = b^2 + c^2 - 2bc\cos A$$

# Practice Exercises

**16.1.** In $\triangle ABC$, $\sin A = \dfrac{1}{2}$, $\sin C = \dfrac{1}{3}$, and $a = 12$. Find the length of side $c$.

**16.2.** In triangle $ABC$, $a = 5$, $b = 7$, and $c = 8$. The measure of $\angle B$ is

    (1) $30°$            (3) $120°$
    (2) $60°$            (4) $150°$

**16.3.** In parallelogram $ABCD$, $AB = 12\,\text{cm}$, $AD = 20\,\text{cm}$, and $m\angle A = 50$.

    *a* Find the length of the longer diagonal of the parallelogram to the *nearest centimeter*.

    *b* Find the area of the parallelogram to the *nearest square centimeter*.

**16.4.** Forces of 40 pounds and 70 pounds act on a body at an angle measuring $60°$. Find the magnitude of the resultant of these forces to the *nearest hundredth of a pound*.

**16.5.** If $m\angle A = 40$, $a = 6$, and $b = 8$, how many distinct triangles can be constructed?

**16.6.** If $m\angle A = 30$, $a = \sqrt{5}$, and $b = 6$, the number of triangles that can be constructed is

(1) 1                    (3) 0
(2) 2                    (4) an infinite number

**16.7.** An airplane traveling at a level altitude of 2050 feet sights the top of a 50-foot tower at an angle of depression of 28° from point $A$. After continuing in level flight to point $B$, the angle of depression to the same tower is 34°. Find, to the *nearest foot*, the distance that the plane traveled from point $A$ to point $B$.

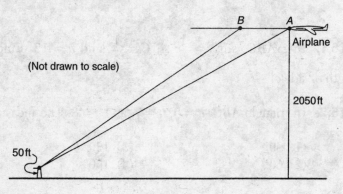

(Not drawn to scale)

B    A
Airplane

2050 ft

50 ft

## Solutions

**16.1.**

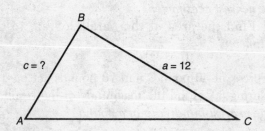

Use the Law of Sines:

$$\frac{\sin A}{a} = \frac{\sin C}{c}$$

Since $\sin A = \dfrac{1}{2}$ and $\sin C = \dfrac{1}{3}$:

$$\frac{\dfrac{1}{2}}{12} = \frac{\dfrac{1}{3}}{c}$$

In a proportion, the product of the means equals the product of the extremes (cross-multiply):

$$\frac{1}{2}c = \frac{1}{3}(12)$$

$$\frac{1}{2}c = 4$$

$$c = 8$$

The length of side $c$ is **8**.

**16.2.**

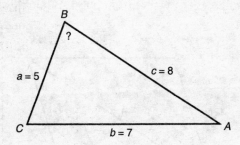

Use the Law of Cosines:

$$b^2 = a^2 + c^2 - 2ac\cos B$$
$$7^2 = 5^2 + 8^2 - 2(5)(8)\cos B$$
$$49 = 25 + 64 - 80\cos B$$
$$49 = 89 - 80\cos B$$
$$49 - 89 = -80\cos B$$
$$-40 = -80\cos B$$
$$\frac{-40}{-80} = \cos B$$
$$\frac{1}{2} = \cos B$$
$$B = 60°$$

The correct choice is **(2)**.

**16.3. a.** The longer diagonal is $\overline{AC}$.

Use the Law of Cosines in $\triangle ADC$:

$$(AC)^2 = (CD)^2 + (AD)^2 - 2(CD)(AD)\cos D$$

Opposite sides of a parallelogram are congruent: $\qquad CD = AB = 12$

Consecutive angles of a parallelogram are supplementary:

$$m\angle D + m\angle A = 180$$
$$m\angle D + 50 = 180$$
$$m\angle D = 180 - 50$$
$$m\angle D = 130$$

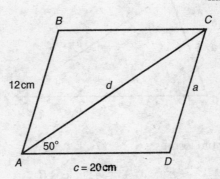

$$(AC)^2 = 12^2 + 20^2 - 2(12)(20)\cos 130°$$

Since 130° is in Quadrant II, where the cosine is negative, $\cos 130° = -\cos 50°$ $= -0.6428$:

$$(AC)^2 = 144 + 400 - 480(-0.6428)$$

$$(AC)^2 = 544 + 308.5440$$

$$(AC)^2 = 852.5440$$

$$AC = \sqrt{852.5440} = 29.1$$

Round off to the *nearest integer*: $\qquad AC = 29$

The length of the longer diagonal is **29 cm** to the *nearest centimeter*.

**b.** The area of a parallelogram equals the product of the lengths of two sides and the sine of the included angle:

$$\text{Area of } \square ABCD = AB \times AD \times \sin A$$
$$= 12 \times 20 \times \sin 50°$$
$\sin 50° \approx 0.7660$:
$$= 240 \times 0.7660$$
$$= 183.84$$

Round off to the *nearest integer*: Area of $\square ABCD = 184$

The area is **184 cm** to the *nearest square centimeter*.

**16.4.** It is given that forces of 40 pounds and 70 pounds act on a body at an angle measuring 60°, as shown in the accompanying diagram. To find the magnitude of the resultant of these forces, complete the parallelogram in which the two forces are adjacent sides and the diagonal of the parallelogram is the resultant.

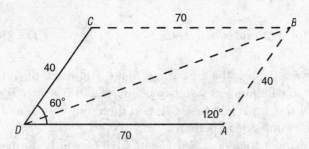

Since consecutive angles of a parallelogram are supplementary, $m\angle A = 180 - 60 = 120$. In $\triangle DAB$, use the Law of Cosines to find $DB$:

$$(DB)^2 = (AD)^2 + (AB)^2 - 2(AD)(AB)\cos A$$
$$= (70)^2 + (40)^2 - 2(70)(40)\cos 120°$$
$$= 4900 + 1600 - 5600(-0.5)$$
$$= 6500 + 2800$$
$$= 9300$$
$$DB = \sqrt{9300} \approx 96.437$$

The magnitude of the resultant force, correct to the *nearest hundredth of a pound*, is **96.44**.

**16.5.** This is the so-called ambiguous case. Simulate the construction of a triangle with the given parts. On one side of $\angle A$, whose measure is 40, mark off the length $AC$, or $b$, equal to 8.

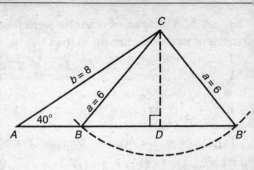

The construction of $\triangle ABC$ must be completed by swinging an arc from $C$ with length $a$ equal to 6, so that it intersects the other side of $\angle A$.

In right $\triangle ADC$:

$$\frac{CD}{AC} = \sin 40°$$

$$\frac{CD}{8} = 0.6428$$

Multiply both sides of the equation by 8:

$$CD = 8(0.6428)$$

$$= 5.1424$$

Since $6 > 5.1424$, the arc swung from $C$ will reach the other side of $\angle A$ and will intersect it in two points, $B$ and $B'$. Since $6 < 8$, the radius of the arc is less than $CA$, and therefore both $B$ and $B'$ will be on the same side of point $A$.

Two distinct triangles, $\triangle ABC$ and $\triangle AB'C$, will be formed, each having the given parts.

**Two** distinct triangles can be constructed.

**16.6.** This is the so-called ambiguous case.

Simulate the construction of a triangle with the given parts. On one side of $\angle A$ (whose measure is 30°), mark off the length $AC = b = 6$. The construction of $\triangle ABC$ must now be completed by swinging an arc from $C$, with length $= \sqrt{5}$. If

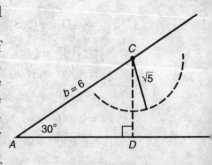

this arc intersects the horizontal side of $\angle A$, the intersection will represent vertex $B$ of the triangle.

If a perpendicular were dropped from $C$ to the opposite side of $\angle A$, it would form $\triangle ACD$, which would be a 30°-60°-90° triangle. In such a triangle, the side opposite 30° is one-half the hypotenuse. Since $AC = 6$, $CD$ would equal 3. Therefore the arc swung from $C$ with a radius of $\sqrt{5}$ could not reach $\overline{AD}$; $\sqrt{5}$ is less than 3 (note that $3^2 = 9$), and the perpendicular from $C$ to $\overline{AD}$, which is the *shortest* distance from $C$ to $\overline{AD}$, is only 3.

No triangle can be constructed with the given dimensions.

The correct choice is (**3**).

**16.7.**

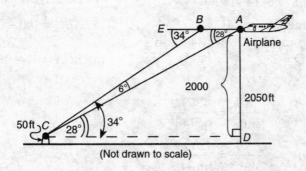

(Not drawn to scale)

Label the top of the tower point as $C$. Draw a line parallel to the ground from $C$ to the vertical segment, labeling the point of intersection $D$ and forming right triangle $ADC$, as shown in the accompanying diagram.

Since the angle of depression from $A$ to $C$ is given as 28°, $m\angle BAC = m\angle ACD = 28$. Also, the angle of depression from $B$ to $C$ is given as 34°, so $m\angle EBC = m\angle BCD = 34$.

Since the distance from $A$ to the ground is given as 2050 feet and the height of the tower as 50 feet, $AD = 2050 - 50 = 2000$.

- In $\triangle ADC$, use the sine ratio to find $AC$:

$$\sin 28° = \frac{AD}{AC}$$

$$0.4695 = \frac{2000}{AC}$$

$$0.4695\,AC = 2000$$

$$AC = \frac{2000}{0.4695} \approx 4259.9$$

- In $\triangle ABC$, since the measures of two angles and a side are known, use the Law of Sines to find $AB$:

$$\frac{\text{side}}{\sin(\angle\text{opposite side})} = \frac{AB}{\sin\angle BCA} = \frac{AC}{\sin\angle ABC}$$

Since $m\angle EBC = 34$:                    $m\angle ABC = 180 - 34 = 146$

Since $m\angle BCD = 34$ and
$m\angle ACD = 28$:                           $m\angle BCA = 34 - 28 = 6$

Hence:

$$\frac{AB}{\sin 6°} = \frac{4259.9}{\sin 146°}$$

$$\frac{AB}{\sin 6°} = \frac{4259.9}{\sin 34°}$$

$$\frac{AB}{0.1045} = \frac{4259.9}{0.5592}$$

$$\frac{AB}{0.1045} = 7617.8$$

$$AB = (0.1045)(7617.8) = 796.06 \approx 796$$

The distance from point $A$ to point $B$, correct to the *nearest foot*, is **796 feet**.

# Glossary of Terms

**abscissa** The $x$-coordinate of a point in the coordinate plane. The abscissa of the point (2,3) is 2.

**absolute value** The absolute value of a number $x$, denoted by $|x|$, is its distance from zero on the number line. Thus, $|x|$ always represents a nonnegative number.

**acute angle** An angle whose measure is less than 90 and greater than 0.

**acute triangle** A triangle that contains three acute angles.

**additive inverse** The opposite of a number. The additive inverse of a number $x$ is $-x$ since $x + (-x) = 0$. The additive inverse of $2 - 3i$ is $-2 + 3i$.

**altitude** A segment that is perpendicular to the side to which it is drawn.

**ambiguous case** The situation in which the measures of two sides and an angle that is not included between the two sides are given. These measures may determine one triangle, two triangles, or no triangles.

**amplitude** The amplitude of a sine or a cosine function of the form $y = a \sin bx$ or $y = a \cos bx$ is $|a|$, which is the maximum height of the graphs of these functions.

**angle** The union of two rays that have the same endpoint.

**angle of depression** The angle formed by a horizontal line of vision and the line of sight when viewing an object beneath the horizontal line of vision.

**angle of elevation** The angle formed by a horizontal line of vision and the line of sight when viewing an object above the horizontal line of vision.

**antilogarithm**  The number whose logarithm is given.

**arc**  A curved part of a circle. If the degree measure of the arc is less than 180, the arc is a **minor arc**. If the degree measure of the arc is greater than 180, the arc is a **major arc**. A **semicircle** is an arc whose degree measure is 180.

**arc cos $x$**  The angle $A$ such that $\cos A = x$ and $0° \leq A \leq 180°$.

**arc sin $x$**  The angle $A$ such that $\sin A = x$ and $-90° \leq A \leq 90°$.

**arc tan $x$**  The angle $A$ such that $\tan A = x$ and $-90° < A < 90°$.

**area of a triangle**  One-half the product of the lengths of any two sides of a triangle and the sine of the included angle.

**associative property**  The mathematical law that states that the order in which three numbers are grouped when they are added or multiplied does *not* matter. For instance, $(2 \times 3) \times 4 = 2 \times (3 \times 4)$.

**asymptote**  A line that a graph approaches but does not intersect as $x$ increases or decreases without bound. The graph of $y = 2^x$ has the negative $x$-axis as an asymptote. The line $x = \dfrac{\pi}{2}$ is an asymptote of the graph of $y = \tan x$.

**axis of symmetry**  For the parabola $y = ax^2 + bx + c$, the vertical line $x = -\dfrac{b}{2a}$.

**Bernoulli experiment**  A probability experiment in which there are exactly two possible outcomes. If one of the possible outcomes is considered a "success" with a probability of $p$, then the remaining outcome is a "failure" with a probability of $1 - p$. The probability of $k$ successes out of $n$ trials is given by the expression ${}_nC_k \, p^k (1-p)^{n-k}$.

**binomial**  A polynomial with two unlike terms, as in $2x + y$.

**binomial theorem**  A formula that tells how to expand a binomial of the form $(a + b)^n$, where $n$ is a positive integer, without performing repeated multiplications. For example, $(a + b)^3 = {}_3C_0 a^3 + {}_3C_1 a^2 b^1 + {}_3C_2 a^1 b^2 + {}_3C_3 b^3$.

**bisect**  To divide into two congruent parts.

**central angle**  The angle whose vertex is at the center of a circle and whose sides are radii.

**change of base formula**  The formula $\log_b N = \dfrac{\log N}{\log b}$ used to change from a logarithm with base $b$ to an equivalent common logarithm expression.

**chord** A line segment whose endpoints are points on a circle.

**circle** The set of points $(x,y)$ in the plane that are a fixed distance $r$ from a given point $(h,k)$ called the *center*. Thus, an equation of a circle is $(x-h)^2 + (y-k)^2 = r^2$.

**coefficient of linear correlation** A number from $-1$ to $+1$, denoted by $r$, that represents the magnitude and direction of a linear relationship, if any, between two sets of data. If a set of data points are closely clustered about a line, then $|r| \approx 1$. The sign of $r$ depends on the sign of the slope of the line about which the data points are clustered.

**collinear points** Points that lie on the same line.

**combination** A selection of objects in which the order of the individual objects is not considered. For example, in selecting a committee of three students from {Alice, Bob, Carol, Dave, Kira}, the combination Alice-Bob-Kira represents the same selection as Bob-Kira-Alice.

**combination formula** The combination of $n$ objects taken $r$ at a time, denoted by $_nC_r$, is given by the formula $_nC_r = \dfrac{n!}{r!(n-r)!}$. For example, the number of different 3-member committees that can be selected from a group of 5 student is

$$_5C_3 = \frac{5!}{3!(5-3)!} = \frac{5 \times 4 \times \cancel{3}!}{\cancel{3}! \times 2 \times 1} = 10.$$

**common logarithm** A logarithm whose base is 10.

**commutative property** The mathematical law that states that the order in which two numbers are added or multiplied does *not* matter. For example, $2 + 3 = 3 + 2$.

**complementary angles** Two angles whose degree measures add up to 90.

**complex fraction** A fraction comprised of other fractions, as in

$$\frac{\frac{1}{2}+1}{\frac{3}{4}}.$$

**complex number** A number that can be written in the form $a + bi$, where $i = \sqrt{-1}$ and $a$ and $b$ are real numbers. The set of real numbers is a subset of the set of complex numbers.

**composition of functions** A function formed by using the output of one function as the input to a second function. The composition of function $f$ followed by function $g$ is the composite function denoted by $g \circ f$, consisting of the set of function values $g(f(x))$, provided $f(x)$ is in the domain of function $g$.

**composition of transformations** A sequence of transformations in which a transformation is applied to the image of another transformation.

**congruent angles (or sides)** Angles (or sides) that have the same measure. The symbol for congruence is $\cong$.

**congruent circles** Circles with congruent radii.

**congruent parts** Pairs of angles or sides that are equal in measure.

**congruent polygons** Two polygons with the same number of sides are congruent if their vertices can be paired so that all corresponding sides have the same length and all corresponding angles have the same degree measure.

**congruent triangles** Two triangles are congruent if any one of the following conditions is true: (1) the three sides of one triangle are congruent to the corresponding sides of the other triangle (SSS $\cong$ SSS); two sides and the included angle of one triangle are congruent to the corresponding parts of the other triangle (SAS $\cong$ SAS); two angles and the included side of one triangle are congruent to the corresponding parts of the other triangle (ASA $\cong$ ASA); two angles and the side opposite one of these angles is congruent to the corresponding parts of the other triangle (AAS $\cong$ AAS).

**conjugate pair** The sum and difference of the same two terms, as in $a + b$ and $a - b$.

**constant** A quantity that is fixed in value. In the equation $y = x + 3$, $x$ and $y$ are variables and 3 is a constant.

**coordinate** The real number that corresponds to the position of a point on the number line.

**coordinate plane** The region formed by a horizontal number line and a vertical number line intersecting at their zero points, called the *origin*.

**cosecant** The reciprocal of the sine function.

**cosine ratio** In a right triangle, the ratio of the length of the leg adjacent to a given acute angle to the length of the hypotenuse. If an angle $\theta$ is in standard position, then $\cos \theta = \dfrac{x}{r}$, where $P(x,y)$ is any point on the terminal side of angle $\theta$ and $r = \sqrt{x^2 + y^2}$.

**cotangent** The reciprocal of the tangent function.

**coterminal angles** Angles in standard position whose terminal sides coincide.

**degree** A unit of angle measurement defined as 1/360 of one complete rotation of a ray about its vertex.

**degree of a monomial** The sum of the exponents of its variable factors. For example, the degree of $3x^4$ is 4, and the degree of $-5xy^2$ is 3 since 1 (the power of $x$) plus 2 (the power of $y$) equals 3.

**degree of a polynomial** The greatest degree of its monomial terms. For example, the degree of $x^2 - 4x + 5$ is 2.

**dependent variable** For a function of the form $y = f(x)$, $y$ is the dependent variable.

**diameter** A chord of a circle that contains the center of the circle.

**dilation** A transformation in which a figure is enlarged or reduced in size according to a given scale factor.

**direct isometry** An isometry that preserves orientation.

**discriminant** In the quadratic formula $x = \dfrac{-b \pm \sqrt{b^2 - 4ac}}{2a}$, the discriminant is the quantity underneath the radical sign, $b^2 - 4ac$. If the discriminant is positive, the two roots are real; if the discriminant is 0, the two roots are equal; and if the discriminant is negative, the two roots are imaginary.

**distributive property of multiplication over addition** For any real numbers $a$, $b$, and $c$, $a(b + c) = ab + ac$ and $(b + c)a = ba + ca$.

**domain of a relation** The set of all possible first members of the ordered pairs that comprise a relation.

**domain of a variable** The set of all possible replacements for a variable. Unless otherwise indicated, the domain of a variable is the largest possible set of real numbers.

**ellipse** An oval-shaped curve, an equation of which is $ax^2 + by^2 = c$, where $a$, $b$, and $c$ have the same sign.

**equation** A statement that two quantities have the same value.

**equilateral triangle** A triangle in which the three sides have the same length.

**equivalent equations** Two equations that have the same solution sets. The equations $2x = 6$ and $x + 1 = 4$ are equivalent since they have the same solution, $x = 3$.

**event** A subset of the set of all possible outcomes of a probability experiment. In flipping a coin the set of all possible outcomes is {head, tail}. One possible event is flipping a head, and another possible event is flipping a tail.

**exponent** In $x^n$, the number $n$ is the exponent and indicates the number of times the base $x$ is used as a factor in a product. Thus, $x^3 = x \cdot x \cdot x$.

**exponential equation** An equation in which the variable appears in an exponent, as in $2^{x+1} = 16$.

**exponential function** A function of the form $y = b^x$, where $b$ is a positive number other than 1.

**exponential regression model** See *regression model*.

**extrapolation** The process of estimating a $y$-value from a table, graph, or equation using a value of $x$ that falls *outside* the range of observed $x$-values.

**extremes** In the proportion $\dfrac{a}{b} = \dfrac{c}{d}$, the terms $a$ and $d$ are the extremes. The extremes are the first and fourth terms of the proportion.

**factor** A number or variable that is being multiplied in a product. A number or variable is a factor of a given product if it divides that product without a nonzero remainder.

**factorial $n$** Denoted by $n!$ and defined for any positive integer $n$ as the product of consecutive integers from $n$ to 1. Thus, $5! = 5 \cdot 4 \cdot 3 \cdot 2 \cdot 1 = 120$.

**factoring** The process by which a number or polynomial is written as the product of two or more terms.

**factoring completely** Factoring a number or polynomial into factors that cannot be factored further.

**FOIL** The rule for multiplying two binomials horizontally by forming the sum of the products of the first terms (F), the outer terms (O), the inner terms (I), and the last terms (L) of each binomial.

**friendly window** The viewing rectangle of a graphing calculator sized so that the cursor moves in "friendly" steps of 0.1, or in multiples of 0.1.

**function** A relation in which no two ordered pairs have the same first member and different second members.

**fundamental principle of counting** If event A can occur in $m$ ways and event B can occur in $n$ ways, then both events can occur in $m$ times $n$ ways.

**glide reflection** The composite of a line reflection and a translation whose direction is parallel to the reflecting line.

**greatest common factor (GCF)** The GCF of two or more monomials is the monomial with the greatest coefficient and the variable factors of the greatest degree that are common to all the given monomials. The GCF of $8a^2b$ and $20ab^2$ is $4ab$.

**horizontal line test** If a horizontal line intersects a graph of a function in, at most, one point, then the graph represents a one-to-one function. If a function is one-to-one, then it has an *inverse* function.

**hyperbola** A curve that consists of two branches, an equation of which is $ax^2 + by^2 = c$, where $a$ and $b$ have opposite signs and $c \neq 0$. A special type of hyperbola, called a rectangular or equilateral hyperbola, has the equation $xy = k$ $(k \neq 0)$ and a graph that is asymptotic to the coordinate axes. A rectangular hyperbola is the graph of an *inverse* variation.

**hypotenuse** The side of a right triangle that is opposite the right angle.

**identity** An equation that is true for all possible replacements of the variable, as in $2x + 3 = 8 - (5 - 2x)$.

**image** In a geometric transformation, the point or figure that corresponds to the original point or figure.

**imaginary number** A number of the form $bi$, where $b$ is a real number and $i$ is the imaginary unit.

**imaginary unit** The number denoted by $i$, where $i = \sqrt{-1}$.

**independent variable** For a function of the form $y = f(x)$, $x$ is the independent variable.

**index** The number $k$ in the expression $\sqrt[k]{x}$ that tells what root of $x$ is to be taken. In a square root radical the index is omitted and is understood to be 2.

**indirect proof** A mathematical proof that shows a statement is true by assuming its opposite is true and then proving the assumption leads to contradiction of a known fact.

**inequality** A sentence that compares two quantities using an inequality relation: < (is less than), ≤ (is less than or equal to), > (is greater than), ≥ (is greater than or equal to), or ≠ (is not equal to).

**inscribed angle** An angle whose vertex is a point on a circle and whose sides are chords.

**integer** A number from the set $\{\ldots,-3,-2,-1,0,1,2,3,\ldots\}$.

**interpolation** The process of estimating a $y$-value from a table, graph, or equation using a value of $x$ that falls *within* the range of observed $x$-values.

**inverse function** The function obtained by interchanging $x$ and $y$ in a one-to-one function and then solving for $y$. The graphs of a function and its inverse are symmetric to the line $y = x$.

**inverse relation** The relation obtained by interchanging the first and second members of each ordered pair of a relation.

**inverse variation** A set of ordered pairs in which the product of the first and second members of each ordered pair is the same nonzero number. Thus, if $y$ varies inversely as $x$, then $xy = k$ or, equivalently, $y = \dfrac{k}{x}$, where $k \neq 0$.

**irrational number** A number that cannot be expressed as the quotient of two integers.

**isometry** A transformation that preserves distance. Reflections, translations, and rotations are isometries. A dilation is not an isometry.

**isosceles triangle** A triangle in which two sides have the same length.

**Law of Cosines** A relationship between the cosine of an angle of a triangle and the lengths of the three sides of the triangle. In $\triangle ABC$,

$$a^2 = b^2 + c^2 - 2bc \cos A$$
$$b^2 = a^2 + c^2 - 2ac \cos B$$
$$c^2 = a^2 + b^2 - 2ab \cos C$$

**Law of Sines** A relationship between two sides of a triangle and the angles opposite these sides. In $\triangle ABC$,

$$\frac{a}{\sin A} = \frac{b}{\sin B} = \frac{c}{\sin C}$$

**least squares regression** A statistical calculation that finds an equation of a line or curve that "best fits" a set of measurement by minimizing the sum of the squares of the vertical distances between the plotted measurements and the fitted line or curve. A graphing calculator has a built-in regression feature that performs the required calculations. See also *regression model*.

**leg of a right triangle** Either of the two sides of a right triangle that include the right angle.

**linear equation** An equation in which the greatest exponent of a variable is 1.

**linear function** A linear equation whose graph is a straight line.

**line reflection** A transformation in which each point $P$ that is not on a line $m$ is paired with a point $P'$ on the opposite side of line $m$ so that line $m$ is the perpendicular bisector of $\overline{PP'}$. If $P$ is on line $m$, then $P'$ coincides with $P$.

**linear regression model** See *regression model*.

**line symmetry** A figure has line symmetry when a line $m$ divides it into two parts such that each part is the reflection of the other part in line $m$.

**logarithm of $x$** An exponent that represents the power to which a given base must be raised to produce a positive number $x$. For example, $\log_2 8 = 3$ because $2^3 = 8$.

**logarithmic function** The function $y = \log_b x$, which is the inverse of the exponential function $y = b^x$, where $b$ is positive and unequal to 1.

**logarithmic regression model** See *regression model*.

**major arc** An arc whose degree measure is greater than 180 and less than 360.

**mapping** A relation in which each member of one set is paired with exactly one member of a second set.

**mean** The mean or average of a set of $n$ data values is the sum of the data values divided by $n$.

**median** The middle value when a set of numbers is arranged in size order. If the set has an even number of values, then the median is the average of the two middle values. For example, the median of 2, 4, 8, 11, and 24 is 8. The median of 7, 11, 23, and 29 is $\dfrac{11 + 23}{2} = 17$.

**median of a triangle** A line segment whose endpoints are a vertex of the triangle and the midpoint of the opposite side.

**minor arc** An arc whose degree measure is less than 180 and greater than 0.

**mode** The data value that occurs most frequently in a given set of data.

**monomial** A number, a variable, or their product.

**multiplicative inverse** The reciprocal of a nonnegative number.

**negative angle** An angle in standard position whose terminal side rotates in a clockwise direction.

**normal curve** A bell-shaped curve that describes a distribution of data values in which approximately 68.2% of the data fall within 1 standard deviation of the mean, 95.4% of the data fall within 2 standard deviations of the mean, and 99.8% of the data fall within 3 standard deviations of the mean.

**obtuse angle** An angle whose degree measure is greater than 90 and less than 180.

**obtuse triangle** A triangle that contains an obtuse angle.

**one-to-one function** A function in which no two ordered pairs have different $x$-values paired with the same $y$-value.

**ordinate** The $y$-coordinate of a point in the coordinate plane. The ordinate of the point (2,3) is 3.

**origin** The zero point on a number line.

**parabola** The U-shaped graph of a quadratic equation in two variables in which either $x$ or $y$ is squared, but not both. A parabola in which $x$ is squared has a vertical axis of symmetry. A parabola in which $y$ is squared has a horizontal axis of symmetry.

**parallelogram** A quadrilateral that has two pairs of parallel sides. In a parallelogram, opposite sides are congruent, consecutive angles are supplementary, and diagonals bisect each other.

**perfect square** A rational number whose square root is also rational.

**period** The length of the smallest interval of $x$ needed for the graph of a cyclic function to repeat itself. The period of a sine or cosine function of the form $y = a \sin bx$ or $y = a \cos bx$ is $\dfrac{2\pi}{|b|}$.

**permutation** An ordered arrangement of objects.

**perpendicular lines** Two lines that intersect to form right angles.

**point symmetry** A figure has point symmetry if after it is rotated 180° (a half-turn) the image coincides with the original figure.

**polynomial** A monomial or the sum or difference of two or more monomials.

**positive angle** An angle in standard position whose terminal side rotates in a counterclockwise direction.

**power regression model** See *regression model*.

**preimage** If under a given transformation, $A'$ is the image of $A$ ($A \rightarrow A'$), then $A$ is the preimage of $A'$.

**prime factorization** The factorization of a polynomial into factors each of which is divisible only by itself or by 1 (or −1).

**probability of an event** A number from 0 to 1 that represents the likelihood that an event will occur. To find the probability of an event, divide the number of equally likely ways the event can occur by the total number of possible outcomes.

**proportion** An equation that states that two ratios are equal. In the proportion $\dfrac{a}{b} = \dfrac{c}{d}$, $a$ and $d$ are called the *extremes* and $b$ and $c$ are called the *means*. In a proportion, the product of the means is equal to the product of the extremes. Thus, $a \times d = b \times c$.

**Pythagorean theorem** The square of the length of the hypotenuse of a right triangle is equal to the sum of the squares of the lengths of the two legs of the right triangle.

**quadrant** One of the four rectangular regions into which the coordinate plane is divided.

**quadrantal angle** An angle in standard position whose terminal side coincides with a coordinate axis. Quadrantal angles include angles whose measures are integer multiples of 90°.

**quadratic equation** An equation that can be put into the standard form $ax^2 + bx + c = 0$, provided $a \neq 0$.

**quadratic formula** The roots of the quadratic equation $ax^2 + bx + c = 0$ are given by the formula $x = \dfrac{-b \pm \sqrt{b^2 - 4ac}}{2a}$, provided $a \neq 0$.

**quadratic function** A function that has the form $y = ax^2 + bx + c$, provided $a \neq 0$.

**radian** The measure of a central angle of a circle that intercepts an arc whose length equals the radius of the circle. The change from degrees to radians, multiply the number of degrees by $\dfrac{\pi}{180°}$. To change from radians to degrees, multiply the number of radians by $\frac{180°}{\pi}$.

**radical equation** An equation in which the variable appears underneath a radical sign as part of the radicand.

**radicand** The expression that appears underneath a radical sign.

**range** The set of all possible second members of the set of ordered pairs that comprise a relation.

**range in data values** The difference between the greatest and the smallest data values.

**rational number** A number that can be written in the form $\frac{a}{b}$, where $a$ and $b$ are integers with $b \neq 0$. Decimals in which a set of digits endlessly repeat, like $2500 \ldots \left( = \frac{1}{4} \right)$ and $0.3333 \ldots \left( = \frac{1}{3} \right)$, represent rational numbers.

**real number** A number that is a member of the set that consists of all rational and irrational numbers.

**rectangle** A parallelogram with four right angles.

**reference angle** When an angle is placed in standard position, the acute angle formed by the terminal side and the $x$-axis.

**reflection** Reflections of points in the coordinate axes are given by the following rules: $r_{x\text{-axis}}(x,y) = (x,-y)$, $r_{y\text{-axis}}(x,y) = (-x,y)$, and $r_{y=x}(x,y) = (y,x)$. To reflect a point in the origin, use the rule $r_{\text{origin}}(x,y) = (-x,-y)$. See also *line reflection*.

**regression model** An equation of the line or curve that is fitted to a set of data by a statistical calculation. A *linear regression* model has the form $y = ax + b$, an *exponential regression* model has the form $y = ab^x$, a *logarithmic regression* model has the form $y = a \ln x + b$, and a *power regression* model has the form $y = ax^b$. The regression feature of a graphing calculator allows you to choose the type of regression model and then calculates the constants $a$ and $b$ for the regression model selected. See also *least squares regression*.

**relation** A set of ordered pairs.

**replacement set** The set of values that a variable can have.

**rhombus** A parallelogram with four congruent sides.

**right angle** An angle whose degree measure is 90. Perpendicular lines intersect at right angles.

**right triangle** A triangle that contains a right angle.

**rotation** A transformation in which a point or figure is turned about a fixed point a given number of degrees. The images of points rotated about the origin through angles that are multiples of $90°$ are given by the following rules: $R_{90°}(x,y) = (-y,x)$, $R_{180°}(x,y) = (-x,-y)$, and $R_{270°}(x,y) = (y,-x)$.

**rotational symmetry** A figure has rotational symmetry if in less than a full rotation of $360°$ the image coincides with the original figure.

**scalene triangle** A triangle in which the three sides have different lengths.

**scatterplot** A graph obtained by representing two sets of data values as a set of ordered pairs and then plotting the ordered pairs in the coordinate plane.

**secant** The reciprocal of the cosine function.

**secant line** A line that intersects a circle in two different points.

**semicircle** An arc whose degree measure is 180.

**sigma** ($\sigma$) The lowercase Greek letter $\sigma$ represents standard deviation.

**sigma** ($\Sigma$) The uppercase Greek letter $\Sigma$ represents the successive summation of terms, as in $\sum_{i=1}^{3} 2^i = 2^1 + 2^2 + 2^3$.

**similar figures** Figures that have the same shape. Two triangles are similar if two pairs of corresponding angles have the same degree measure. When two triangles are similar, the lengths of corresponding sides are proportional.

**sine ratio** In a right triangle, the ratio of the length of the leg that is opposite a given acute angle to the length of the hypotenuse. If an angle $\theta$ is in standard position, then $\sin\theta = \dfrac{y}{r}$, where $P(x,y)$ is any point on the terminal side of angle $\theta$ and $r = \sqrt{x^2 + y^2}$.

**square** A parallelogram with four right angles and four congruent sides.

**standard deviation** A statistic that measures how spread out numerical data are from the mean.

**standard position** The position of an angle whose vertex is fixed at the origin and whose initial side coincides with the positive $x$-axis. The side of the angle that rotates is called the *terminal side*. A counterclockwise rotation of the terminal side of an angle represents a positive angle, and a clockwise rotation of the terminal side of an angle produces a negative angle. If an angle $\theta$ is in standard position and $P(x,y)$ is any point on the terminal side of angle $\theta$, where $r = \sqrt{x^2 + y^2}$, then for all values of $\theta$ for which the trigonometric functions are defined:

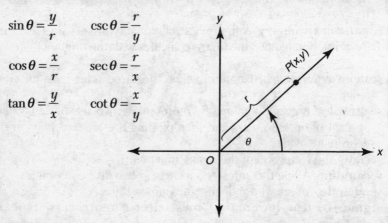

$$\sin\theta = \frac{y}{r} \qquad \csc\theta = \frac{r}{y}$$

$$\cos\theta = \frac{x}{r} \qquad \sec\theta = \frac{r}{x}$$

$$\tan\theta = \frac{y}{x} \qquad \cot\theta = \frac{x}{y}$$

The algebraic signs of the trigonometric functions depend on the signs of $x$ and $y$ in the particular quadrant in which the terminal side of angle $\theta$ is located.

**standard window** The viewing rectangle of a graphing calculator sized so that each coordinate axis has 10 tick marks on either side of the origin.

**supplementary angles** Two angles whose degree measures add up to 180.

**tangent line** A line that intersects a circle in exactly one point.

**tangent ratio** In a right triangle, the ratio of the length of the leg that is opposite a given acute angle to the length of the leg adjacent to that angle. If an angle $\theta$ is in standard position, then $\tan\theta = \frac{y}{r}$, where $P(x,y)$ is any point on the terminal side of angle $\theta$.

**terminal side** The side of an angle in standard position that rotates about the origin while the initial side of the angle remains fixed on the positive $x$-axis.

**theorem** A generalization in mathematics that can be proved.

**transformation** A change made to a figure by using a given rule to map each of its points onto one and only one point in a set of image points.

**translation** A transformation in which each point of a figure is moved the same distance and in the same direction. The transformation $T_{h,k}$ slides a point $h$ units horizontally and $k$ units vertically. Thus, $T_{h,k}(x, y) = (x + h, y + k)$.

**trapezoid** A quadrilateral in which exactly one pair of sides are parallel. The nonparallel sides are called *legs*.

**trinomial** A polynomial with three terms, as in $x^2 - 3x + 7$.

**unit circle** A circle with a radius of 1.

**vertex** The highest or lowest point of a parabola. The $x$-coordinate of the vertex of the parabola $y = ax^2 + bx + c$ is $x = -\dfrac{b}{2a}$. In the vertex form of the equation of the parabola $y = a(x - h)^2 + k$, the vertex is $(h,k)$.

**vertical angles** Opposite pairs of angles formed when two lines intersect.

**vertical line test** If no vertical line intersects a graph in more than one point, the graph represents a function.

**zero of a function** Any value of the variable for which the function evaluates to 0. Each $x$-intercept of the graph of a function, if any, represents a zero of the function.

# Regents Examinations, Answers, and Self-Analysis Charts

# Examination August 2001
## Math B

## PART I

**Answer all questions in this part. Each correct answer will receive 2 credits. No partial credit will be allowed. Circle your answer choice.** [40]

1 Which relation is *not* a function?
   (1) $y = 2x + 4$      (3) $x = 3y - 2$
   (2) $y = x^2 - 4x + 3$      (4) $x = y^2 + 2x - 3$      1 _____

2 The solution set of $|3x + 2| < 1$ contains
   (1) only negative real numbers
   (2) only positive real numbers
   (3) both positive and negative real numbers
   (4) no real numbers      2 _____

3 In the accompanying diagram, cabins $B$ and $G$ are located on the shore of a circular lake, and cabin $L$ is located near the lake. Point $D$ is a dock on the lake shore and is collinear with cabins $B$ and $L$. The road between cabins $G$ and $L$ is 8 miles long and is tangent to the lake. The path between cabin $L$ and dock $D$ is 4 miles long.

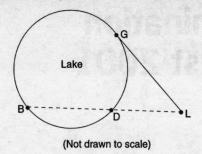

(Not drawn to scale)

What is the length, in miles, of $\overline{BD}$ ?
(1) 24                    (3) 8
(2) 12                    (4) 4          3____

4 The solution set of the equation $\sqrt{x+6} = x$ is
(1) {−2,3}                (3) {3}
(2) {−2}                  (4) { }        4____

5 Which transformation is a direct isometry?
(1) $D_2$                 (3) $r_{y\text{-axis}}$
(2) $D_{-2}$             (4) $T_{2,5}$    5____

6 The roots of the equation $x^2 - 3x - 2 = 0$ are
(1) real, rational, and equal
(2) real, rational, and unequal
(3) real, irrational, and unequal
(4) imaginary                            6____

7 The new corporate logo created by the design engi-
neers at Magic Motors is shown in the accompanying diagram.

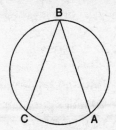

If chords $\overline{BA}$ and $\overline{BC}$ are congruent and $m\widehat{BC}$ = 140, what is $m\angle B$?

(1) 40            (3) 140

(2) 80            (4) 280      7____

8 At Mogul's Ski Resort, the beginner's slope is inclined at an angle of 12.3°, while the advanced slope is inclined at an angle of 26.4°. If Rudy skis 1,000 meters down the advanced slope while Valerie skis the same distance on the beginner's slope, how much longer was the horizontal distance that Valerie covered?

(1) 81.3 m        (3) 895.7 m

(2) 231.6 m       (4) 977.0 m      8____

9 A regular hexagon is inscribed in a circle. What is the ratio of the length of a side of the hexagon to the minor arc that it intercepts?

(1) $\dfrac{\pi}{6}$            (3) $\dfrac{3}{\pi}$

(2) $\dfrac{3}{6}$            (4) $\dfrac{6}{\pi}$      9____

10  If $\log 5 = a$, then $\log 250$ can be expressed as
   (1) $50a$                     (3) $10 + 2a$
   (2) $2a + 1$                   (4) $25a$                     10 _____

11  On a trip, a student drove 40 miles per hour for 2
   hours and then drove 30 miles per hour for 3 hours.
   What is the student's average rate of speed, in miles
   per hour, for the whole trip?
   (1) 34                        (3) 36
   (2) 35                        (4) 37                        11 _____

12  A ball is thrown straight up at an initial velocity of
   54 feet per second. The height of the ball $t$ seconds
   after it is thrown is given by the formula $h(t) = 54t -
   12t^2$. How many seconds after the ball is thrown
   will it return to the ground?
   (1) 9.2                       (3) 4.5
   (2) 6                         (4) 4                         12 _____

13  What is the period of the function $y = 5 \sin 3x$?
   (1) 5                         (3) 3
   (2) $\dfrac{2\pi}{5}$         (4) $\dfrac{2\pi}{3}$         13 _____

14  A cellular telephone company has two plans. Plan $A$
   charges \$11 a month and \$0.21 per minute. Plan $B$
   charges \$20 a month and \$0.10 per minute. After
   how much time, to the *nearest minute*, will the cost
   of plan $A$ be equal to the cost of plan $B$?
   (1) 1 hr 22 min               (3) 81 hr 8 min
   (2) 1 hr 36 min               (4) 81 hr 48 min              14 _____

15 The graph of f($x$) is shown in the accompanying diagram.

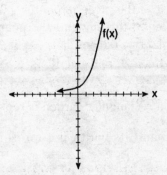

Which graph represents f($x$)$_{r_{x\text{-axis}} \circ r_{y\text{-axis}}}$?

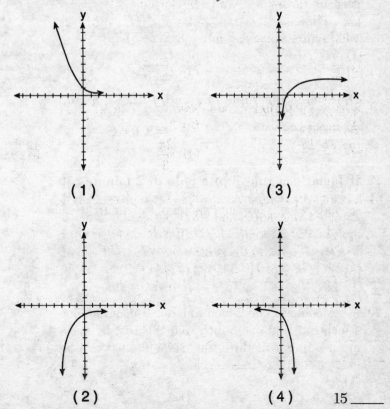

(1)

(3)

(2)

(4)

15 _____

16 A wedge-shaped piece is cut from a circular pizza. The radius of the pizza is 6 inches. The rounded edge of the crust of the piece measures 4.2 inches. To the *nearest tenth*, the angle of the pointed end of the piece of pizza, in radians, is

(1) 0.7          (3) 7.0

(2) 1.4          (4) 25.2          16 _____

17 If the length of a rectangular garden is represented by $\dfrac{x^2+2x}{x^2+2x-15}$ and its width is represented by $\dfrac{2x-6}{2x+4}$, which expression represents the area of the garden?

(1) $x$          (3) $\dfrac{x^2+2x}{2(x+5)}$

(2) $x+5$          (4) $\dfrac{x}{x+5}$          17 _____

18 Determine the value of $x$ and $y$ if $2^y = 8^x$ and $3^y = 3^{x+4}$.

(1) $x = 6, y = 2$          (3) $x = 2, y = 6$

(2) $x = -2, y = -6$          (4) $x = y$          18 _____

19 If Jamar can run $\dfrac{3}{5}$ of a mile in 2 minutes 30 seconds, what is his rate in miles per minute?

(1) $\dfrac{4}{5}$          (3) $3\dfrac{1}{10}$

(2) $\dfrac{6}{25}$          (4) $4\dfrac{1}{6}$          19 _____

20 A box contains one 2-inch rod, one 3-inch rod, one 4-inch rod, and one 5-inch rod. What is the maximum number of different triangles that can be made using these rods as sides?

(1) 1          (3) 3

(2) 2          (4) 4          20 _____

# PART II

**Answer all questions in this part. Each correct answer will receive 2 credits. Clearly indicate the necessary steps, including appropriate formula substitutions, diagrams, graphs, charts, etc. For all questions in this part, a correct numerical answer with no work shown will receive only 1 credit.**   [12]

21 If the sine of an angle is $\frac{3}{5}$ and the angle is *not* in Quadrant I, what is the value of the cosine of the angle?

22 Show that the product of $a + bi$ and its conjugate is a real number.

23 The price per person to rent a limousine for a prom varies inversely as the number of passengers. If five people rent the limousine, the cost is $70 each. How many people are renting the limousine when the cost *per couple* is $87.50?

24 The accompanying diagram shows a semicircular arch over a street that has a radius of 14 feet. A banner is attached to the arch at points $A$ and $B$, such that $AE = EB = 5$ feet. How many feet above the ground are these points of attachment for the banner?

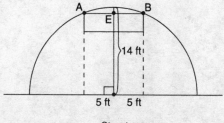

Street

25 Working by herself, Mary requires 16 minutes more than Antoine to solve a mathematics problem. Working together, Mary and Antoine can solve the problem in 6 minutes. If this situation is represented by the equation $\dfrac{6}{t} + \dfrac{6}{t+16} = 1$, where $t$ represents the number of minutes Antoine works alone to solve the problem, how many minutes will it take Antoine to solve the problem if he works by himself?

26 If $\sin x = \dfrac{4}{5}$, where $0° < x < 90°$, find the value of $\cos(x + 180°)$.

# PART III

**Answer all questions in this part. Each correct answer will receive 4 credits. Clearly indicate the necessary steps, including appropriate formula substitutions, diagrams, graphs, charts, etc. For all questions in this part, a correct numerical answer with no work shown will receive only 1 credit.   [24]**

27 The times of average monthly sunrise, as shown in the accompanying diagram, over the course of a 12-month interval can be modeled by the equation $y = A\cos(Bx) + D$. Determine the values of $A$, $B$, and $D$, and explain how you arrived at your values.

**Time of Average Monthly Sunrise**

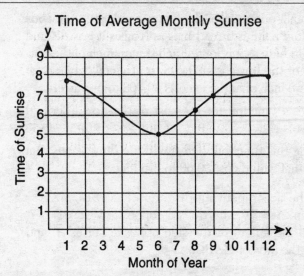

28  As shown in the accompanying diagram, a circular target with a radius of 9 inches has a bull's-eye that has a radius of 3 inches. If five arrows randomly hit the target, what is the probability that *at least* four hit the bull's-eye?

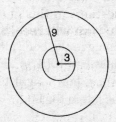

29  Twenty high school students took an examination and received the following scores:

70, 60, 75, 68, 85, 86, 78, 72, 82, 88, 88, 73, 74, 79, 86, 82, 90, 92, 93, 73

Determine what percent of the students scored within one standard deviation of the mean. Do the results of the examination approximate a normal distribution? Justify your answer.

30  A small, open-top packing box, similar to a shoebox without a lid, is three times as long as it is wide, and half as high as it is long. Each square inch of the bottom of the box costs $0.008 to produce, while each square inch of any side costs $0.003 to produce.

Write a function for the cost of the box described above.

Using this function, determine the dimensions of a box that would cost $0.69 to produce.

31  In the accompanying diagram of $\triangle ABC$, $m\angle A = 65$, $m\angle B = 70$, and the side opposite vertex $B$ is 7. Find the length of the side opposite vertex $A$, and find the area of $\triangle ABC$.

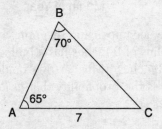

32  The amount $A$, in milligrams, of a 10-milligram dose of a drug remaining in the body after $t$ hours is given by the formula $A = 10(0.8)^t$. Find, to the *nearest tenth of an hour*, how long it takes for half of the drug dose to be left in the body.

## PART IV

**Answer all questions in this part. Each correct answer will receive 6 credits. Clearly indicate the necessary steps, including appropriate formula substitutions, diagrams, graphs, charts, etc. For all questions in this part, a correct numerical answer with no work shown will receive only 1 credit.**  [12]

33  The availability of leaded gasoline in New York State is decreasing, as shown in the accompanying table.

| Year | 1984 | 1988 | 1992 | 1996 | 2000 |
|------|------|------|------|------|------|
| **Gallons Available** (in thousands) | 150 | 124 | 104 | 76 | 50 |

Determine a linear relationship for $x$ (years) versus $y$ (gallons available), based on the data given. The data should be entered using the year and gallons available (in thousands), such as (1984,150).

If this relationship continues, determine the number of gallons of leaded gasoline available in New York State in the year 2005.

If this relationship continues, during what year will leaded gasoline first become unavailable in New York State?

34 Given: $A(1,6)$, $B(7,9)$, $C(13,6)$, and $D(3,1)$
Prove: $ABCD$ is a trapezoid. [*The use of the accompanying grid is optional.*]

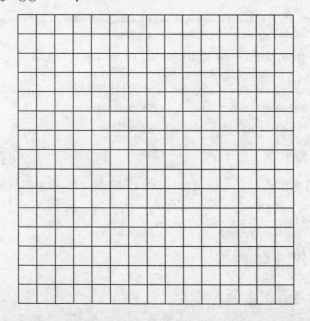

# Answers
# August 2001
## Math B

## Answer Key

## PART I

| | | | |
|---|---|---|---|
| **1.** (4) | **6.** (3) | **11.** (1) | **16.** (1) |
| **2.** (1) | **7.** (1) | **12.** (3) | **17.** (4) |
| **3.** (2) | **8.** (1) | **13.** (4) | **18.** (3) |
| **4.** (3) | **9.** (3) | **14.** (1) | **19.** (2) |
| **5.** (4) | **10.** (2) | **15.** (2) | **20.** (3) |

**For parts II, III, and IV, see Answers Explained section for computations and methodology to support the given answers.**

## PART II

**21.** $-\dfrac{4}{5}$

**22.** a real number

**23.** 8

**24.** $\sqrt{171}$ feet

**25.** 8 minutes

**26.** $-\dfrac{3}{5}$

## PART III

**27.** $A = \dfrac{3}{2}, B = \pi$

**28.** $\dfrac{41}{59049}$

**29.** a normal distribution

**30.** $W = \sqrt{11.5}; L = 3\sqrt{11.5}; H = \dfrac{3}{2}\sqrt{11.5}$

**31.** 6.75 and 16.71

**32.** 3.1 hours

## PART IV

**33.** 2008

**34.** $ABCD$ is a trapezoid

# Answers Explained

## PART I

**1.** In a function each value of $x$ is associated with only one value of $y$. Choice (4) is the relation $x = y^2 + 2x - 3$. In this equation the variable $y$ appears only as a $y^2$ term. Hence positive and negative values of $y$ are associated with the same value of $x$. For example, if $x = 2$, then $y = \pm 1$.

Since there are two values of $y$ associated with the same value of $x$, the relation $x = y^2 + 2x - 3$ is *not* a function.

The correct choice is **(4)**.

**2.** The given inequality is:                                 $|3x + 2| < 1$
Rewrite the inequality without an absolute value:          $-1 < 3x + 2 < 1$
Solve the inequality:                                      $-3 < 3x < -1$

$$-1 < x < -\frac{1}{3}$$

The solution set contains only negative real numbers.

The correct choice is **(1)**.

**3.** In the accompanying diagram it is given that $\overline{LG}$ is tangent to the circle.

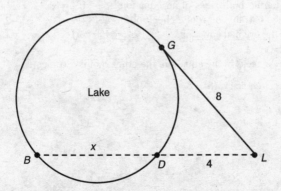

When a tangent and a secant are drawn to a circle from the same external point, the tangent segment is the geometric mean of the external segment of the secant and the entire secant.

It is also given that $LG = 8$ and $LD = 4$. Let $BD = x$.

In any proportion the product of the means equals the product of the extremes.

Solve the equation for $x$, the length of $\overline{BD}$:

$$\frac{LD}{LG} = \frac{LG}{LB}$$
$$\frac{4}{8} = \frac{8}{x+4}$$
$$4(x+4) = 64$$
$$4x + 16 = 64$$
$$4x = 48$$
$$x = 12$$

The correct choice is (**2**).

**4.** The given equation is:

$$\sqrt{x+6} = x$$

To solve an equation involving a radical, isolate the radical and square both sides:

$$x + 6 = x^2$$

To solve the quadratic equation, rewrite it in standard form with 0 on one side and all other terms on the other side:

$$0 = x^2 - x - 6$$

Factor the quadratic trinomial as the product of two binomials:

$$0 = (x-3)(x+2)$$

If the product of two quantities is 0, then at least one of them is 0. Solve the two equations for $x$:

$$0 = x - 3 \text{ or } 0 = x + 2$$
$$x = 3 \quad \text{ or } \quad x = -2$$

Since squaring both sides of an equation can introduce extraneous roots, check both solutions in the original equation:

If $x = 3$;
$\sqrt{3+6} = 3$?
$\sqrt{9} = 3$✓

If $x = -2$:
$\sqrt{-2+6} = -2$?
$\sqrt{4} = 2 \neq -2$No!

Only $x = 3$ checks in the equation; the other root is extraneous.

The correct choice is (**3**).

**5.** A direct isometry is a transformation that preserves both size and orientation. Consider each of the choices.

- Choice (1): $D_2$ is a dilation, which does not preserve size. Choice (1) is not correct.
- Choice (2): $D_{-2}$ is also a dilation. Choice (2) is not correct.
- Choice (3): $r_{y\text{-axis}}$ is a reflection, which does not preserve orientation. Choice (3) is not correct.
- Choice (4): $T_{2,5}$ is a translation, which preserves both size and orientation. Choice (4) is a direct isometry.

The correct choice is **(4)**.

**6.** The given quadratic equation $x^2 - 3x - 2 = 0$ is in standard form $ax^2 + bx + c = 0$, with $a = 1$, $b = -3$, and $c = -2$. The nature of the roots can be determined by evaluating the discriminant: $b^2 - 4ac = (-3)^2 - 4(1)(-2) = 17$.

The discriminant is the expression under the square root in the quadratic formula. Because 17 is positive, the roots are real. Because 17 is not a perfect square, the roots are irrational. Because 17 is not 0, the roots are unequal.

The correct choice is **(3)**.

**7.** In the accompanying diagram it is given that $\overline{BA} \cong \overline{BC}$ and $\text{m}\widehat{BC} = 140$.

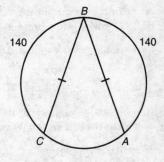

If two chords of a circle are congruent, their arcs are congruent. Hence $\text{m}\widehat{BA} = 140$.

In a circle the sum of the measures of all the arcs must be 360. Hence $\text{m}\widehat{CA} = 360 - 2(140) = 80$.

In a circle an inscribed angle is measured by one-half its intercepted arc. Hence $\text{m} \angle B = \frac{1}{2}\text{m}\widehat{CA} = \frac{1}{2}(80) = 40$.

The correct choice is **(1)**.

**8.** It is given that two ski slopes each 1000 meters long are inclined at 26.4° and 12.3° as shown in the accompanying diagrams.

Let $R$ and $V$ represent the horizontal distances covered by the two skiers.

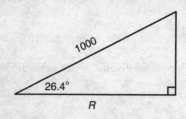

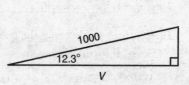

To find $R$ and $V$ use the cosine function and your calculator:

$$\cos 26.4 = \frac{R}{1000} \qquad \cos 12.3 = \frac{V}{1000}$$
$$R = 1000 \cos 26.4 \qquad V = 1000 \cos 12.3$$
$$R \approx 895.71 \qquad V \approx 977.05$$

Valerie's distance is approximately $977.05 - 895.71 = 81.34$ meters longer.

The correct choice is (**1**).

**9.** In the accompanying diagram regular hexagon $ABCDEF$ is inscribed in circle $O$. Since $\overset{\frown}{AB}$ represents one-sixth of the circle, its length is one-sixth of the circumference: length of $\overset{\frown}{AB} = \frac{1}{6}(2\pi r) = \frac{\pi r}{3}$.

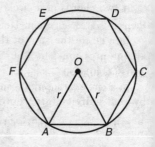

In isosceles $\triangle AOB$, $m \angle AOB = \frac{1}{6}(360) = 60$.

Hence $\triangle AOB$ is equilateral, and $AB = r$.

The ratio of the lengths of side $\overline{AB}$ to arc $\overset{\frown}{AB}$ is $\frac{r}{\pi r/3} = \frac{3}{\pi}$.

The correct choice is (**3**).

**10.** The given logarithm is:

$\log 250$

Rewrite 250 using factors of 5 and 10:

$\log(5^2 \cdot 10)$

The logarithm of a product equals the sum of the logarithms of the factors:

$\log 5^2 + \log 10$

The logarithm of a number raised to a power equals the power times the logarithm of the number, and $\log 10 = 1$. Also, it is given that $\log 5 = a$.

$2 \log 5 \ + \log 10$

$2a + 1$

The correct choice is (**2**).

**11.** During the 2 hours the student drove 40 miles per hour, she went a distance of $2(40) = 80$ miles. During the 3 hours the student drove 30 miles per hour, she went a distance of $3(30) = 90$ miles. For the whole 5-hour trip, the student traveled $80 + 90 = 170$ miles. Hence the student's average speed for the whole trip was $170 \div 5 = 34$ miles per hour.

The correct choice is **(1)**.

**12.** The ball's height above the ground $t$ seconds after it is thrown is given by the formula $h(t) = 54t - 12t^2$.

To determine when it will return to the ground, set its height equal to 0 and solve the quadratic equation for $t$:

$$0 = 54t - 12t^2$$
$$0 = 6t(9 - 2t)$$
$$0 = 6t \text{ or } 0 = 9 - 2t$$
$$0 = t \text{ or } 4.5 = t$$

The ball was on the ground at $t = 0$ (when it was thrown) and returns at $t = 4.5$ seconds.

The correct choice is **(3)**.

**13.** The given function is $y = 5 \sin 3x$. Since any function of the form $y = \sin t$ completes one cycle during $0 \le t \le 2\pi$, the given function completes one cycle during $0 \le 3x \le 2\pi$, or $\dfrac{2\pi}{3}$. Hence the period of the given equation is $\dfrac{2\pi}{3}$.

The correct choice is **(4)**.

**14.** It is given that Plan A costs \$11 plus \$0.21 per minute and Plan B costs \$20 plus \$0.10 per minute.

Let $m$ = the number of minutes of cellular phone calls. Then the costs of the two plans can be represented by:

$$A = 11 + 0.21m \text{ and } B = 20 + 0.10m$$

The costs will be equal when:

$$11 + 0.21m = 20 + 0.10m$$

Solve for $m$:

$$0.11m = 9$$
$$m \approx 81.8$$

The cost of the two plans will be equal after approximately 82 minutes, or 1 hour and 22 minutes.

The correct choice is **(1)**.

**15.** To find the graph of $f(x)_{r_{x\text{-axis}} \circ r_{y\text{-axis}}}$, first reflect the graph of $f(x)$ across the $y$-axis, finding its image $f^*(x)$, as shown in the accompanying diagram.

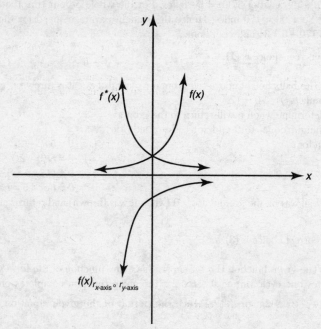

Now reflect the image $f^*(x)$ across the $x$-axis. This new image is the graph of $f(x)_{r_{x\text{-axis}} \circ r_{y\text{-axis}}}$.

The correct choice is (**2**).

**16.** As shown in the accompanying diagram, the radius of the pizza is 6 inches and the length of the arc of crust is 4.2 inches.

In a circle of radius 6 a central angle of 1 radian intersects an arc 6 inches long. Hence this central angle is $\dfrac{4.2}{6} = 0.7$ radians.

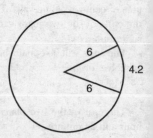

The correct choice is (**1**).

**17.** The area of a rectangle is the product of its length and width:

$$\frac{x^2 + 2x}{x^2 + 2x - 15} \cdot \frac{2x - 6}{2x + 4}$$

To express this area in simplest form, first factor the expression in each numerator and denominator. In the first numerator the terms have $x$ as a common factor. In the first denominator the quadratic trinomial can be factored into the product of two binomials. In the second fraction the numerator and denominator each have 2 as a common factor:

$$\frac{x(x + 2)}{(x + 5)(x - 3)} \cdot \frac{2(x - 3)}{2x + 4}$$

Cancel factors that appear in both a numerator and a denominator, indicating that their quotient is 1:

$$\frac{\overset{1}{x\cancel{(x+2)}}}{\underset{1}{(x+5)\cancel{(x-3)}}} \cdot \frac{\overset{1}{\cancel{2}}\overset{1}{\cancel{(x-3)}}}{\underset{1}{\cancel{2}}\underset{1}{\cancel{(x+2)}}}$$

In simplest form, the area of the garden is $\dfrac{x}{x + 5}$.

The correct choice is (**4**).

**18.** The given equations are:
Rewrite the first equation using a common base:
Set the exponents equal:
Solve the two equations simultaneously by substituting $3x$ for $y$ in the second equation:

$2^y = 8^x$ and $3^y = 3^{x+4}$

$2^y = 2^{3x}$ and $3^y = 3^{x+4}$

$y = 3x$ and $y = x + 4$

$3x = x + 4$

$2x = 4$

$x = 2$

$y = 3x = 6$

The correct choice is (**3**).

**19.** Expressed as a fraction, 2 minutes 30 seconds $= \dfrac{5}{2}$ minutes.

When Jamar runs $\dfrac{3}{5}$ of a mile in $\dfrac{5}{2}$ minutes, his rate is $\dfrac{3}{5} \div \dfrac{5}{2} = \dfrac{3}{5} \cdot \dfrac{2}{5} = \dfrac{6}{25}$ miles per minute.

The correct choice is (**2**).

**20.** In any triangle the sum of the lengths of any two sides must be greater than the length of the third side. We are given rods of lengths 2, 3, 4, and 5. Consider the four ways to choose three of these rods.

- $\{2,3,4\}$: Since $2 + 3 > 4$, these three rods can form a triangle.
- $\{2,3,5\}$: Since $2 + 3 = 5$, these three rods *cannot* form a triangle.
- $\{2,4,5\}$: Since $2 + 4 > 5$, these three rods can form a triangle.
- $\{3,4,5\}$: Since $3 + 4 > 5$, these three rods can form a triangle.

There are three different triangles that can be made using these rods as sides.

The correct choice is **(3)**.

# PART II

**21.** It is given that the sine of an angle is $\frac{3}{5}$ and the angle is *not* in Quadrant I. Since the sine of an angle is positive in Quadrant I or Quadrant II, the given angle must be in Quadrant II, as shown in the accompanying diagram.

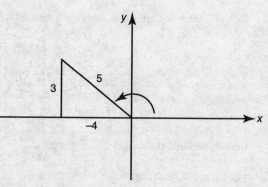

The sine of an angle is the ratio of the vertical leg of its reference triangle to the hypotenuse, so label these sides 3 and 5, respectively. By the Pythagorean Theorem, the length of the third side of this reference triangle is 4. Since the triangle is in Quadrant II, label the horizontal side −4, as shown in the diagram.

The cosine of an angle is the ratio of the horizontal leg of its reference triangle to the hypotenuse. Hence the cosine of the given angle is $\frac{-4}{5}$.

The cosine of the angle is $-\frac{4}{5}$.

**22.** In a complex number $a + bi$ the coefficients $a$ and $b$ are real numbers. The conjugate of $a + bi$ is the complex number $a - bi$.

The product of the number and its conjugate is:
$$(a + bi)(a - bi) = a^2 - b^2 i^2$$
$$= a^2 - b^2 (-1)$$
$$= a^2 + b^2$$

Since $a$ and $b$ are both real numbers, $a^2 + b^2$ is also a real number. Therefore, the product of $a + bi$ and its conjugate is a real number.

**23.** It is given that the price per person to rent a limousine varies inversely as the number of passengers. Let $c$ = the cost per person and $p$ = the number of passengers.

If two variables vary inversely, then their product is a constant. Call this constant $k$:

$$cp = k$$

It is given that when 5 people rent the limousine, the cost is $70 each. Substitute $p = 5$ and $c = 70$ to find the value of $k$:

$$70 \cdot 5 = k$$
$$350 = k$$

When the cost *per couple* is $87.50, the cost per person is $87.50 \div 2 = \$43.75$; so substitute $c = 43.75$ and find the value of $p$:

$$43.75\,p = 350$$
$$p = 8$$

The number of people renting the limousine is **8**.

**24.** It is given that the radius of the arch is 14 feet and that $AE = EB = 5$, as shown in the accompanying diagram.

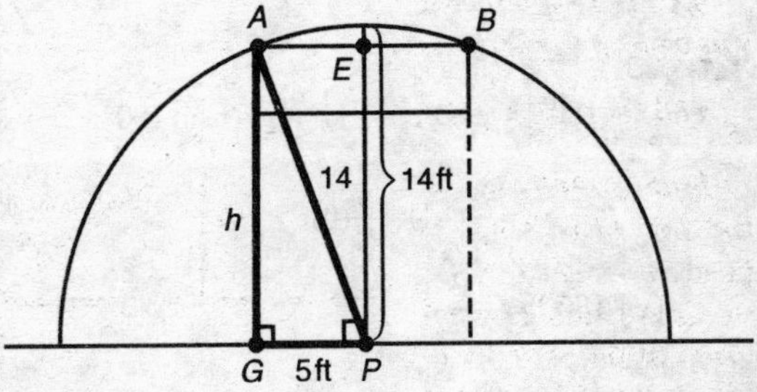

Draw radius $\overline{PA}$ as shown, forming right triangle $PGA$ with $PA = 14$ and $PG = 5$. Let $h$ represent the distance of the banner above the ground.

Find $h$ using the Pythagorean Theorem:

$$5^2 + h^2 = 14^2$$
$$25 + h^2 = 196$$
$$h^2 = 171$$
$$h = \sqrt{171}$$

The points of attachment are $\sqrt{171}$ **feet** above the ground.

**25.** The given equation is:

$$\frac{6}{t} + \frac{6}{t+16} = 1$$

To solve an equation involving fractions, multiply each term by the common denominator. The common denominator of $t$ and $t + 16$ is $t(t + 16)$:

$$\frac{6}{t}t(t+16) + \frac{6}{t+16}t(t+16) = 1t(t+16)$$

Simplify:

$$6(t+16) + 6t = t(t+1)$$
$$6t + 96 + 6t = t^2 + 16t$$
$$96 + 12t = t^2 + 16t$$

To solve the quadratic equation, rewrite it in standard form with 0 on one side and all other terms on the other side:

$$0 = t^2 + 4t - 96$$

Factor the quadratic trinomial as the product of two binomials:

$$0 = (t + 12)(t - 8)$$

If the product of two quantities is 0, then at least one of them is 0. Solve the two equations for $t$:

$$t + 12 = 0 \text{ or } t - 8 = 0$$
$$t = -12 \text{ or } t = 8$$

It will take Antoine **8 minutes** to solve the problem if he works by himself.

**26.** It is given that $\sin x = \dfrac{4}{5}$, where $0° < x < 90°$, as shown in the accompanying diagram. Since the reference triangle is a 3-4-5 right triangle, $\cos x = \dfrac{3}{5}$.

$$\cos(x + 180°) = \cos x \cos 180° - \sin x \sin 180°$$

$$= \frac{3}{5}(-1) - \frac{4}{5}(0)$$

$$\cos(x + 180°) = -\frac{3}{5}$$

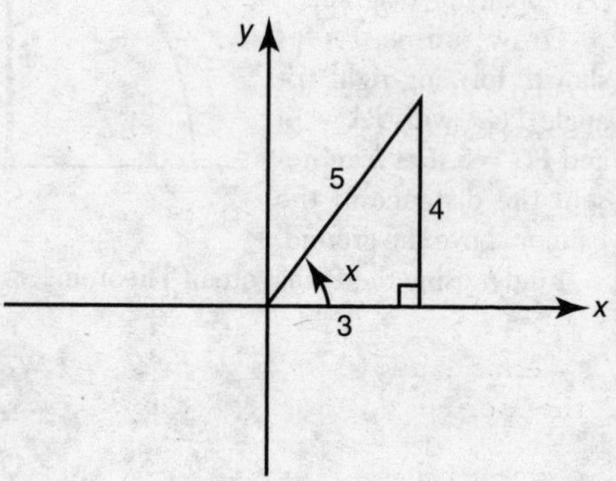

# PART III

**27.** In an equation of the form $y = A\cos(Bx) + D$, $A$ determines the amplitude of the curve, $B$ determines the period of the curve, and $D$ represents a vertical translation of the curve.

In the accompanying diagram, the earliest sunrise is at 5 A.M. and the latest at 8 A.M. The amplitude of the curve is $\dfrac{1}{2}(8-5) = \dfrac{3}{2}$. Hence $A = \dfrac{3}{2}$.

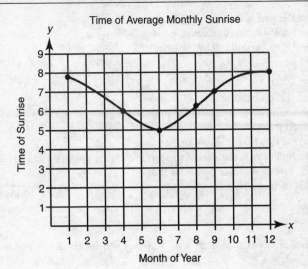

Time of Average Monthly Sunrise

A cosine curve with amplitude $\dfrac{3}{2}$ would reach a maximum value of $\dfrac{3}{2}$. Since the given curve reaches a maximum of 8, it must be translated upward $8 - \dfrac{3}{2} = \dfrac{13}{2}$ units. Hence $D = \dfrac{13}{2}$.

Any function of the form $y = \cos Bx$ completes one cycle when $Bx = 2\pi$. Since the period of the given curve is 12 months, the sunrise cycle is complete when $x = 12$. Solving $B \cdot 12 = 2\pi$ yields $B = \dfrac{\pi}{6}$.

The values are $A = \dfrac{3}{2}$, $B = \dfrac{\pi}{6}$, and $D = \dfrac{13}{2}$.

**28.** The area of the bull's-eye shown in the accompanying diagram is $\pi \cdot 3^2 = 9\pi$ square inches, and the area of the entire target is $\pi \cdot 9^2 = 81\pi$ square inches. Hence the probability that an arrow hitting the target at random will hit the bull's-eye is $\dfrac{9\pi}{81\pi} = \dfrac{1}{9}$.

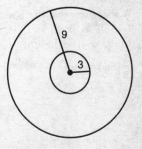

To find the probability that exactly 4 out of 5 arrows will hit the bull's-eye, evaluate the probability formula $_nC_xp^x(1-p)^{n-x}$, where $n=5$, $x=4$, and $p=\dfrac{1}{9}$:

$$P(x=4) =\ _5C_4\left(\frac{1}{9}\right)^4\left(\frac{8}{9}\right)^1 = \frac{40}{59049}$$

To find the probability that all 5 arrows will hit the bull's-eye, evaluate the probability formula using $x=5$:

The probability that at least 4 of 5 arrows will hit the bull's-eye is the sum of the probability that exactly 4 out of 5 will hit it and the probability that all 5 will.

$$P(x=5) =\ _5C_5\left(\frac{1}{9}\right)^5\left(\frac{8}{9}\right)^0 = \frac{1}{59049}$$

$$P(x \ge 4) = P(x=4) + P(x=5)$$

$$= \frac{40}{59049} + \frac{1}{59049}$$

$$= \frac{41}{59049}$$

The probability that at least 4 of 5 arrows will hit the bull's-eye is $\dfrac{\mathbf{41}}{\mathbf{59049}}$.

**29.** Find the mean and the standard deviation of the given set of 20 exam scores using your calculator: $\bar{x} = 79.7$ and $s \approx 8.7$.

One standard deviation above the mean is a score of $79.7 + 8.7 = 88.4$.

One standard deviation below the mean is a score of $79.7 - 8.7 = 71.0$.

Of the 20 exam scores, 14 are between 71 and 88.4. Hence 70% of the exam scores are within one standard deviation of the mean.

Since 68% of a normal distribution is found within one standard deviation of the mean, the results of this examination approximate a normal distribution.

**30.** Let $w$ = the width of the open-topped box shown in the accompanying diagram. It is given that the box is three times as long as it is wide; so the length is $3w$. It is also given that the box is half as high as it is long; so the height is $\frac{1}{2}(3w) = \frac{3}{2}w$.

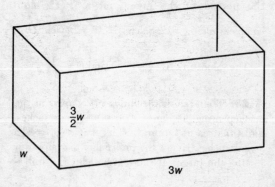

- Producing the bottom costs $0.008 per square inch, or $0.008(w)(3w)$.

- Producing each end costs $0.003 per square inch, or $0.003(3w)\left(\dfrac{3}{2}w\right)$.

- Producing the front or back costs $0.003 per square inch, or $0.003(3w)\left(\dfrac{3}{2}w\right)$.

The total cost, $C$, of producing the box with one bottom, two ends, one front, and one back is given by:

$$C = 0.008(w)(3w) + 2\left[0.003(w)\left(\frac{3}{2}w\right)\right] + 2\left[0.003(3w)\left(\frac{3}{2}w\right)\right]$$

Simplify: $\qquad\qquad\qquad\qquad\qquad\qquad\qquad\qquad C = 0.06w^2$

To determine the dimensions of a box costing
$0.69, let $C = 0.69$ and solve: $\qquad\qquad\qquad\qquad 0.69 = 0.06w^2$

$$w = \sqrt{11.5}$$

The dimensions of the box are width = $\sqrt{11.5}$, length = $3\sqrt{11.5}$, and height = $\dfrac{3}{2}\sqrt{11.5}$.

**31.** In the accompanying diagram of $\triangle ABC$ it is given that $m\angle A = 65$, $m\angle B = 70$, and $AC = 7$.

Find the length of the side opposite $A$ using the Law of Sines and your calculator:

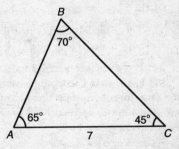

$$\frac{\sin 65}{BC} = \frac{\sin 70}{7}$$

$$BC\ \sin 70 = 7 \sin 65$$

$$BC = \frac{7 \sin 65}{\sin 70}$$

$$BC \approx 6.75$$

The area of a triangle is one-half the product of any two sides and the sine of the included angle.

$$m\angle C = 180 - 70 - 65$$
$$m\angle C = 45$$
$$\text{Area} = \frac{1}{2}(7)(6.75)\sin 45$$
$$\text{Area} \approx 16.71$$

The length of the side opposite vertex $A$ is **6.75**, and the area of $\triangle ABC$ is **16.71**.

**32.** The given formula is: $\qquad A = 10(0.8)^t$

To find how long it will be until half of the 10-milligram drug dose is left, substitute $A = 5$: $\qquad 5 = 10(0.8)^t$

Solve the equation for $t$:
$$0.5 = 0.8^t$$
$$\log 0.5 = t \log 0.8$$
$$t = \frac{\log 0.5}{\log 0.8}$$

Evaluate using your calculator: $\qquad\qquad t \approx 3.106$

To the nearest tenth of an hour, half the dose will be left in **3.1 hours**.

# PART IV

**33.** To determine the linear relationship, enter the given data into your calculator with years as $x$ and thousands of gallons available as $y$.

Find the equation of the line of best fit: $\qquad y = -6.2x + 12451.2$

To determine the number of gallons available in 2005, substitute $x = 2005$ and evaluate $y$: $\qquad y = -6.2(2005) + 12451.2 = 20.2$

The linear model predicts that if this relationship continues, there will be **20200 gallons** of leaded gasoline available in New York in 2005.

To determine when leaded gasoline will become unavailable, substitute $y = 0$ and solve the equation for $x$:
$$0 = -6.2x + 12451.2$$
$$6.2x = 12451.2$$
$$x = 2008.258$$

The linear model predicts that if this relationship continues, leaded gasoline will become unavailable in New York during **2008**.

**34.** It is given that quadrilateral $ABCD$ has vertices $A(1,6)$, $B(7,9)$, $C(13,6)$, and $D(3,1)$, as shown in the accompanying diagram. To prove $ABCD$ is a trapezoid, use the slope formula to show that two of the sides are parallel and the other two are not. A quadrilateral with only one pair of opposite sides parallel is a trapezoid.

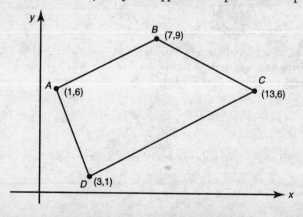

STEP 1: Two lines are parallel if their slopes are equal. To show that $\overline{AB}\|\overline{DC}$, find the slope of each segment.

The slope, $m$, of a line through the points $(x_1,y_1)$ and $(x_2,y_2)$ can be determined using the formula $m = \dfrac{y_2 - y_1}{x_2 - x_1}$.

- To find the slope of $\overline{AB}$, let $(x_1,y_1) = A(1,6)$ and $(x_2,y_2) = B(7,9)$:

$$m = \frac{y_2 - y_1}{x_2 - x_1} = \frac{9-6}{7-1} = \frac{3}{6} = \frac{1}{2}$$

- To find the slope of $\overline{DC}$, let $(x_1,y_1) = D(3,1)$ and $(x_2,y_2) = C(13,6)$:

$$m = \frac{y_2 - y_1}{x_2 - x_1} = \frac{6-1}{13-3} = \frac{5}{10} = \frac{1}{2}$$

Since the slope of $\overline{AB}$ equals the slope of $\overline{DC}$, $\overline{AB}\|\overline{DC}$.

STEP 2: Two lines are not parallel if their slopes are not equal. To show that $\overline{AD}$ is not parallel to $\overline{CB}$, find the slope of each segment.

- To find the slope of $\overline{AD}$, let $(x_1,y_1) = A(1,6)$ and $(x_2,y_2) = D(3,1)$:

$$m = \frac{y_2 - y_1}{x_2 - x_1} = \frac{1-6}{3-1} = -\frac{5}{2}$$

- To find the slope of $\overline{CB}$, let $(x_1,y_1) = C(13,6)$ and $(x_2,y_2) = B(7,9)$:

$$m = \frac{y_2 - y_1}{x_2 - x_1} = \frac{9-6}{7-13} = \frac{3}{-6} = -\frac{1}{2}$$

Since the slope of $\overline{AD}$ does not equal the slope of $\overline{CB}$, $\overline{AD}$ is not parallel to $\overline{CB}$.

CONCLUSION: $ABCD$ is a trapezoid because two of the sides are parallel and the other two are not. A quadrilateral with only one pair of opposite sides parallel is a trapezoid.

| Topic | Question Numbers | Number of Points | Your Points |
|---|---|---|---|
| 1. Properties of Real Numbers | — | — | |
| 2. Sequences | — | — | |
| 3. Complex Numbers | — | — | |
| 4. Inequalities, Absolute Value | 2, 20 | $2 + 2 = 4$ | |
| 5. Algebraic Expressions, Fractions | 17, 25 | $2 + 2 = 4$ | |
| 6. Exponents (zero, negative, rational, scientific notation) | — | — | |
| 7. Radical Expressions | 4 | 2 | |
| 8. Quadratic Equations (factors, formula, discriminant) | 6, 12 | $2 + 2 = 4$ | |
| 9. Systems of Equations | 18 | 2 | |
| 10. Functions (graphs, domain, range, roots) | 1 | 2 | |
| 11. Inverse Functions, Composition | — | — | |
| 12. Linear Functions (including rates of change) | 11, 14, 19 | $2 + 2 + 2 = 6$ | |
| 13. Parabolas (max/min, axis of symmetry) | — | — | |
| 14. Hyperbolas (including inverse variation) | 23 | 2 | |
| 15. Ellipses | — | — | |
| 16. Exponents and Logarithms | 10, 32 | $2 + 2 = 4$ | |
| 17. Trig (circular functions, unit circle, radians) | 13, 16, 21, 27 | $2 + 2 + 2 + 4 = 10$ | |
| 18. Trig Equations and Identities | 8, 26 | $2 + 2 = 4$ | |
| 19. Laws of Sin and Cos, Pythagorean Theorem, ambiguous case | 24, 31 | $2 + 4 = 6$ | |
| 20. Coordinate Geometry (slope, distance, midpoint, circle) | — | — | |
| 21. Transformations, Symmetry | 5, 15 | $2 + 2 = 4$ | |
| 22. Circle Geometry | 3, 7, 9 | $2 + 2 + 2 = 6$ | |
| 23. Congruence, Similarity, Proportions | — | — | |
| 24. Probability (including Bernoulli events) | 28 | 4 | |
| 25. Normal Curve | 29 | 2 | |
| 26. Statistics (center, spread, summation notation) | 29 | 2 | |
| 27. Correlation, Modeling, Prediction | 33 | 6 | |
| 28. Algebraic Proofs, Word Problems | 22, 30 | $2 + 4 = 6$ | |
| 29. Geometric Proofs | — | — | |
| 30. Coordinate Geometry Proofs | 34 | 6 | |

Total Raw Score = _____

**NOTE:** Find your point total in the "Raw Score" column, and then locate the corresponding "Scaled Score." This scaled score is your **final exam score**.

| Raw Score | Scaled Score | Raw Score | Scaled Score | Raw Score | Scaled Score |
|---|---|---|---|---|---|
| 88 | 100 | 58 | 83 | 28 | 50 |
| 87 | 99 | 57 | 82 | 27 | 48 |
| 86 | 99 | 56 | 81 | 26 | 47 |
| 85 | 99 | 55 | 80 | 25 | 45 |
| 84 | 98 | 54 | 79 | 24 | 44 |
| 83 | 98 | 53 | 78 | 23 | 42 |
| 82 | 98 | 52 | 77 | 22 | 41 |
| 81 | 97 | 51 | 77 | 21 | 39 |
| 80 | 97 | 50 | 76 | 20 | 37 |
| 79 | 96 | 49 | 75 | 19 | 36 |
| 78 | 96 | 48 | 74 | 18 | 34 |
| 77 | 95 | 47 | 73 | 17 | 32 |
| 76 | 95 | 46 | 72 | 16 | 31 |
| 75 | 94 | 45 | 71 | 15 | 29 |
| 74 | 94 | 44 | 69 | 14 | 27 |
| 73 | 93 | 43 | 68 | 13 | 26 |
| 72 | 92 | 42 | 67 | 12 | 24 |
| 71 | 92 | 41 | 66 | 11 | 22 |
| 70 | 91 | 40 | 65 | 10 | 20 |
| 69 | 91 | 39 | 64 | 9 | 18 |
| 68 | 90 | 38 | 63 | 8 | 16 |
| 67 | 89 | 37 | 61 | 7 | 14 |
| 66 | 89 | 36 | 60 | 6 | 12 |
| 65 | 88 | 35 | 59 | 5 | 10 |
| 64 | 87 | 34 | 58 | 4 | 8 |
| 63 | 86 | 33 | 56 | 3 | 6 |
| 62 | 86 | 32 | 55 | 2 | 4 |
| 61 | 85 | 31 | 54 | 1 | 2 |
| 60 | 84 | 30 | 52 | 0 | 0 |
| 59 | 83 | 29 | 51 | | |

# Examination January 2002
## Math B

**FORMULAS**

### Area of Triangle

$$K = \frac{1}{2}ab \sin C$$

### Function of the Sum of Two Angles

$$\sin (A + B) = \sin A \cos B + \cos A \sin B$$
$$\cos (A + B) = \cos A \cos B - \sin A \sin B$$

### Function of the Difference of Two Angles

$$\sin (A - B) = \sin A \cos B - \cos A \sin B$$
$$\cos (A - B) = \cos A \cos B + \sin A \sin B$$

### Law of Sines

$$\frac{a}{\sin A} = \frac{b}{\sin B} = \frac{c}{\sin C}$$

### Law of Cosines

$$a^2 = b^2 + c^2 - 2bc \cos A$$

### Functions of the Double Angle

$$\sin 2A = 2 \sin A \cos A$$
$$\cos 2A = \cos^2 A - \sin^2 A$$
$$\cos 2A = 2 \cos^2 A - 1$$
$$\cos 2A = 1 - 2 \sin^2 A$$

## Functions of the Half Angle

$$\sin\frac{1}{2}A = \pm\sqrt{\frac{1 - \cos A}{2}}$$

$$\cos\frac{1}{2}A = \pm\sqrt{\frac{1 + \cos A}{2}}$$

## Normal Curve
## Standard Deviation

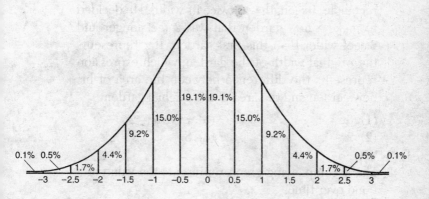

# PART I

Answer all questions in this part. Each correct answer will receive 2 credits. No partial credit will be allowed. Record your answers in the spaces provided.   [40]

1 The roots of a quadratic equation are real, rational, and equal when the discriminant is

(1) –2                          (3) 0
(2) 2                           (4) 4                          1 _____

2 Chad had a garden that was in the shape of a rectangle. Its length was twice its width. He decided to make a new garden that was 2 feet longer and 2 feet wider than this first garden. If $x$ represents the original width of the garden, which expression represents the difference between the area of his new garden and the area of the original garden?

(1) $6x + 4$                    (3) $x^2 + 3x + 2$
(2) $2x^2$                      (4) 8                          2 _____

3 The accompanying graph represents the value of a bond over time.

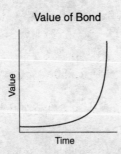

Which type of function does this graph best model?

(1) trigonometric               (3) quadratic
(2) logarithmic                 (4) exponential                3 _____

4 An object that weighs 2 pounds is suspended in a liquid. When the object is depressed 3 feet from its equilibrium point, it will oscillate according to the formula $x = 3 \cos (8t)$, where $t$ is the number of seconds after the object is released. How many seconds are in the period of oscillation?

(1) $\dfrac{\pi}{4}$                      (3) 3

(2) $\pi$                       (4) $2\pi$           4 \_\_\_\_

5 If $\theta$ is an angle in standard position and its terminal side passes through the point $\left( \dfrac{1}{2}, \dfrac{\sqrt{3}}{2} \right)$ on a unit circle, a possible value of $\theta$ is

(1) 30°                   (3) 120°

(2) 60°                   (4) 150°         5 \_\_\_\_

6 The expression $\dfrac{\dfrac{a}{b} - \dfrac{b}{a}}{\dfrac{1}{a} + \dfrac{1}{b}}$ is equivalent to

(1) $a + b$              (3) $ab$

(2) $a - b$              (4) $\dfrac{a-b}{ab}$         6 \_\_\_\_

7 If $f(x) = 5x^2$ and $g(x) = \sqrt{2x}$, what is the value of $(f \circ g)(8)$?

(1) $8\sqrt{10}$            (3) 80

(2) 16                  (4) 1,280         7 \_\_\_\_

8 Which expression is *not* equivalent to $\log_b 36$?

(1) $6 \log_b 2$         (3) $2 \log_b 6$

(2) $\log_b 9 + \log_b 4$     (4) $\log_b 72 - \log_b 2$      8 \_\_\_\_

9 If a function is defined by the equation $y = 3x + 2$, which equation defines the inverse of this function?

(1) $x = \frac{1}{3}y + \frac{1}{2}$          (3) $y = \frac{1}{3}x - \frac{2}{3}$

(2) $y = \frac{1}{3}x + \frac{1}{2}$          (4) $y = -3x - 2$      9 \_\_\_\_

10 Which transformation is *not* an isometry?

(1) $r_{y=x}$          (3) $T_{3,6}$

(2) $R_{0,90°}$          (4) $D_2$      10 \_\_\_\_

11 Which relation is a function?

(1) $x = 4$          (3) $y = \sin x$

(2) $x = y^2 + 1$          (4) $x^2 + y^2 = 16$      11 \_\_\_\_

12 In $\triangle ABC$, $m\angle A = 33$, $a = 12$, and $b = 15$. What is $m\angle B$ to the *nearest degree*?

(1) 41          (3) 44

(2) 43          (4) 48      12 \_\_\_\_

13 The accompanying diagram represents circular pond $O$ with docks located at points $A$ and $B$. From a cabin located at $C$, two sightings are taken that determine an angle of 30° for tangents $\overrightarrow{CA}$ and $\overrightarrow{CB}$.

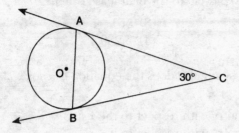

What is m$\angle CAB$?

(1) 30                        (3) 75
(2) 60                        (4) 150                        13 _____

14 The accompanying diagram shows a section of a sound wave as displayed on an oscilloscope.

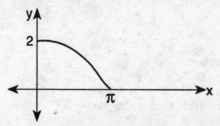

Which equation could represent this graph?

(1) $y = 2 \cos \frac{x}{2}$          (3) $y = \frac{1}{2} \cos 2x$

(2) $y = 2 \sin \frac{x}{2}$          (4) $y = \frac{1}{2} \sin \frac{\pi}{2} x$          14 _____

15 Every time the pedals go through a 360° rotation on a certain bicycle, the tires rotate three times. If the tires are 24 inches in diameter, what is the minimum number of complete rotations of the pedals needed for the bicycle to travel at least 1 mile?

(1) 12            (3) 561

(2) 281          (4) 5,280          15 \_\_\_\_

16 Which type of symmetry does the equation $y = \cos x$ have?

(1) line symmetry with respect to the $x$-axis

(2) line symmetry with respect to $y = x$

(3) point symmetry with respect to the origin

(4) point symmetry with respect to $\left(\dfrac{\pi}{2}, 0\right)$      16 \_\_\_\_

17 The value of $\left(\dfrac{3^0}{27^{\frac{2}{3}}}\right)^{-1}$ is

(1) $-9$           (3) $-\dfrac{1}{9}$

(2) $9$            (4) $\dfrac{1}{9}$          17 \_\_\_\_

18 What is the domain of $h(x) = \sqrt{x^2 - 4x - 5}$ ?

(1) $\{x | x \geq 1 \text{ or } x \leq -5\}$    (3) $\{x | -1 \leq x \leq 5\}$

(2) $\{x | x \geq 5 \text{ or } x \leq -1\}$    (4) $\{x | -5 \leq x \leq 1\}$      18 \_\_\_\_

19 The expression $(-1 + i)^3$ is equivalent to

(1) $-3i$          (3) $-1 - i$

(2) $-2 - 2i$        (4) $2 + 2i$          19 \_\_\_\_

20 The revenue, $R(x)$, from selling $x$ units of a product is represented by the equation $R(x) = 35x$, while the total cost, $C(x)$, of making $x$ units of the product is represented by the equation $C(x) = 20x + 500$. The total profit, $P(x)$, is represented by the equation $P(x) = R(x) - C(x)$. For the values of $R(x)$ and $C(x)$ given above, what is $P(x)$?

(1) $15x$            (3) $15x - 500$

(2) $15x + 500$       (4) $10x + 100$       20 _____

# PART II

**Answer all questions in this part. Each correct answer will receive 2 credits. Clearly indicate the necessary steps, including appropriate formula substitutions, diagrams, graphs, charts, etc. For all questions in this part, a correct numerical answer with no work shown will receive only 1 credit.** [12]

21 Explain how a person can determine if a set of data represents inverse variation and give an example using a table of values.

22 Solve for $x$ in simplest $a + bi$ form: $x^2 + 8x + 25 = 0$

23 A ball is rolling in a circular path that has a radius of 10 inches, as shown in the accompanying diagram. What distance has the ball rolled when the subtended arc is 54°? Express your answer to the *nearest hundredth of an inch.*

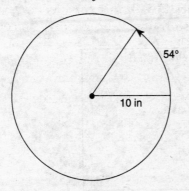

24 A rectangle is said to have a golden ratio when $\frac{w}{h} = \frac{h}{w - h}$, where $w$ represents width and $h$ represents height. When $w = 3$, between which two consecutive integers will $h$ lie?

25 The accompanying diagram shows the floor plan
for a kitchen. The owners plan to carpet all of
the kitchen except the "work space," which is
represented by scalene triangle *ABC*. Find the area
of this work space to the *nearest tenth of a square
foot*.

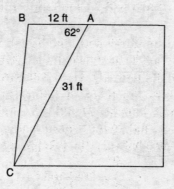

26 A set of normally distributed student test scores
has a mean of 80 and a standard deviation of 4.
Determine the probability that a randomly selected
score will be between 74 and 82.

## PART III

**Answer all questions in this part. Each correct answer will receive 4 credits. Clearly indicate the necessary steps, including appropriate formula substitutions, diagrams, graphs, charts, etc. For all questions in this part, a correct numerical answer with no work shown will receive only 1 credit.** [24]

27 Two straight roads, Elm Street and Pine Street, intersect creating a 40° angle, as shown in the accompanying diagram. John's house ($J$) is on Elm Street and is 3.2 miles from the point of intersection. Mary's house ($M$) is on Pine Street and is 5.6 miles from the intersection. Find, to the *nearest tenth of a mile*, the direct distance between the two houses.

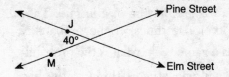

28  At the local video rental store, José rents two movies and three games for a total of $15.50. At the same time, Meg rents three movies and one game for a total of $12.05. How much money is needed to rent a combination of one game and one movie?

29  Team $A$ and team $B$ are playing in a league. They will play each other five times. If the probability that team $A$ wins a game is $\frac{1}{3}$, what is the probability that team $A$ will win *at least* three of the five games?

30 Depreciation (the decline in cash value) on a car can be determined by the formula $V = C(1 - r)^t$, where $V$ is the value of the car after $t$ years, $C$ is the original cost, and $r$ is the rate of depreciation. If a car's cost, when new, is $15,000, the rate of depreciation is 30%, and the value of the car now is $3,000, how old is the car to the *nearest tenth of a year*?

31 When a baseball is hit by a batter, the height of the ball, $h(t)$, at time $t$, $t \geq 0$, is determined by the equation $h(t) = -16t^2 + 64t + 4$. For which interval of time is the height of the ball greater than or equal to 52 feet?

32 *a* On the accompanying grid, graph the equation $2y = 2x^2 - 4$ in the interval $-3 \leq x \leq 3$ and label it *a*.

  *b* On the same grid, sketch the image of *a* under $T_{5,-2} \circ r_{x\text{-}axis}$ and label it *b*.

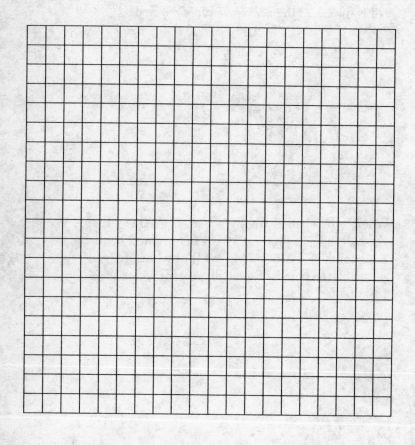

## PART IV

Answer all questions in this part. Each correct answer will receive 6 credits. Clearly indicate the necessary steps, including appropriate formula substitutions, diagrams, graphs, charts, etc. For all questions in this part, a correct numerical answer with no work shown will receive only 1 credit.   [12]

33 Prove that the diagonals of a parallelogram bisect each other.

34 Two different tests were designed to measure understanding of a topic. The two tests were given to ten students with the following results:

| Test $x$ | 75 | 78 | 88 | 92 | 95 | 67 | 58 | 72 | 74 | 81 |
|---|---|---|---|---|---|---|---|---|---|---|
| Test $y$ | 81 | 73 | 85 | 88 | 89 | 73 | 66 | 75 | 70 | 78 |

Construct a scatter plot for these scores, and then write an equation for the line of best fit (round slope and intercept to the *nearest hundredth*).

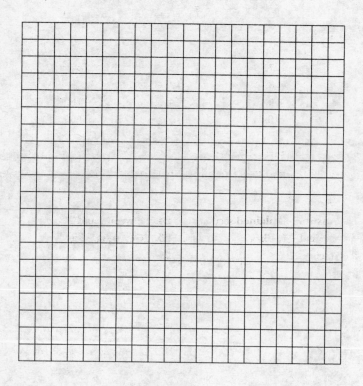

Find the correlation coefficient.

Predict the score, to the *nearest integer*, on test $y$ for student who scored 87 on test $x$.

# Answers
# January 2002
## Math B

### Answer Key

## PART I

| | | | |
|---|---|---|---|
| **1.** 3 | **6.** 2 | **11.** 3 | **16.** 4 |
| **2.** 1 | **7.** 3 | **12.** 2 | **17.** 2 |
| **3.** 4 | **8.** 1 | **13.** 3 | **18.** 2 |
| **4.** 1 | **9.** 3 | **14.** 1 | **19.** 4 |
| **5.** 2 | **10.** 4 | **15.** 2 | **20.** 3 |

**For parts II, III, and IV, see Answers Explained section for computations and methodology to support the given answers.**

## PART II

**21.** See **Answers Explained** section

**22.** $-4 + 3i$ and $-4 - 3i$

**23.** 9.42 inches

**24.** between 1 and 2

**25.** 164.2 square feet

**26.** 62.4%

## PART III

**27.** 3.8 miles

**28.** $6.15

**29.** $\dfrac{51}{243}$

**30.** 4.5 years

**31.** $1 \leq t \leq 3$

**32.** See **Answers Explained** section

## PART IV

**33.** See **Answers Explained** section

**34.** 83

# Answers Explained

## PART I

**1.** The discriminant, $b^2 - 4ac$, appears under the square root in the quadratic formula. It is given that the roots are real; therefore the discriminant must be nonnegative. It is also given that the roots are rational; therefore the discriminant must be a perfect square. And it is given that the roots are equal; therefore the discriminant must equal zero.

The correct choice is **(3)**.

**2.** It is given that the length of a rectangle is twice its width. If $x$ represents the width, then $2x$ represents the length, as shown in the accompanying diagram.

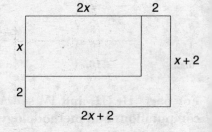

Let $x + 2$ and $2x + 2$ represent the dimensions of the larger rectangle that is 2 feet wider and two feet longer than the original.

The area of a rectangle is the product of its length and width. Express the difference in areas in terms of $x$:

Simplify the expression:

$$(2x + 2)(x + 2) - (2x)(x)$$
$$2x^2 + 4x + 2x + 4 - 2x^2$$
$$6x + 4$$

The correct choice is **(1)**.

**3.** The given graph shows a function that is concave upward with a positive $y$-intercept. The function increases more quickly as $x$ increases, and there are no apparent symmetries. These are all properties of exponential functions.

The correct choice is **(4)**.

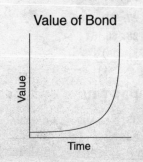

### Value of Bond

**4.** The given function is $x = 3 \cos (8t)$. Since any function of the form $y = \cos x$ completes one cycle during $0 \le x \le 2\pi$, the given function completes one cycle during $0 \le 8t \le 2\pi$, or $0 \le t \le \dfrac{\pi}{4}$. Hence the period of the given equation is $\dfrac{\pi}{4}$.

The correct choice is (**1**).

**5.** When an angle $\theta$ is in standard position in a unit circle, as shown in the accompanying diagram, its terminal side passes through the point $(\cos \theta, \sin \theta)$.

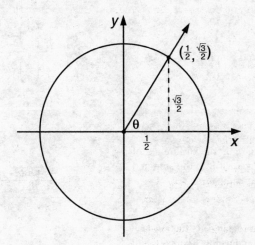

Since the given angle passes through the point $\left( \dfrac{1}{2}, \dfrac{\sqrt{3}}{2} \right)$, $\cos \theta = \dfrac{1}{2}$. Hence $\theta = 60°$.

The correct choice is (**2**).

**6.** The given expression is:

$$\frac{\dfrac{a}{b}-\dfrac{b}{a}}{\dfrac{1}{a}+\dfrac{1}{b}}$$

To simplify this complex fraction, multiply the numerator and denominator by the common denominator of the fraction terms, $ab$:

$$\frac{ab\left(\dfrac{a}{b}-\dfrac{b}{a}\right)}{ab\left(\dfrac{1}{a}+\dfrac{1}{b}\right)}$$

Distribute and simplify:

$$\frac{ab\left(\dfrac{a}{b}\right)-ab\left(\dfrac{b}{a}\right)}{ab\left(\dfrac{1}{a}\right)+ab\left(\dfrac{1}{b}\right)}$$

$$\frac{a^2-b^2}{b+a}$$

The numerator is the difference of two squares. Factor it into the sum and difference of the same terms:

$$\frac{\cancel{(a+b)}(a-b)}{\cancel{(b+a)}}$$

Since $a + b = b + a$, the fraction may be reduced by canceling the common factor:

$$(a-b)$$

The correct choice is (2).

**7.** The given functions are $f(x) = 5x^2$ and $g(x) = \sqrt{2x}$. To evaluate $f \circ g(8)$, first find $g(8)$: $g(8) = \sqrt{2 \cdot 8} = 4$.

Then $f \circ g(8) = f(g(8)) = f(4) = 5 \cdot 4^2 = 80$.

The correct choice is (3).

**8.** Since the logarithm of a number raised to a power equals the power times the logarithm of the number, $6 \log_b 2 = \log_b 2^6 = \log_b 64$. This is *not* equivalent to $\log_b 36$.

The correct choice is **(1)**.

**9.** The given function is: $\qquad\qquad\qquad\qquad\qquad\qquad y = 3x + 2$

To find an equation of the inverse of this

function, first interchange $x$ and $y$: $\qquad\qquad\qquad\qquad x = 3y + 2$

Solve this equation for $y$: $\qquad\qquad\qquad\qquad\quad x - 2 = 3y$

$$\frac{1}{3}x - \frac{2}{3} = y$$

The correct choice is **(3)**.

**10.** An isometry is a transformation that preserves size. Consider each of the choices.

- Choice (1): $r_{y = x}$ is a reflection, which preserves size. Choice (1) is an isometry.
- Choice (2): $R_{0,90°}$ is a rotation, which preserves size. Choice (2) is an isometry.
- Choice (3): $T_{3,6}$ is a translation, which preserves size. Choice (3) is an isometry.
- Choice (4): $D_2$ is a dilation by a factor of 2, which does *not* preserve size. Choice (4) is *not* an isometry.

The correct choice is **(4)**.

**11.** In a function each value of $x$ is associated with only one value of $y$. Consider each of the choices.

- Choice (1) is the vertical line $x = 4$. This value of $x$ is associated with every value of $y$. Choice (1) is *not* a function.
- Choice (2) is the parabola $x = y^2 + 1$. In this equation the variable $y$ appears only as a $y^2$-term. Hence, positive and negative values of $y$ are associated with the same value of $x$. For example, if $x = 2$, then $y = \pm1$. Choice (2) is *not* a function.
- Choice (3) is the sine curve $y = \sin x$. For each angle $x$ there is only one value of $\sin x$. Choice (3) *is* a function.
- Choice (4) is the circle $x^2 + y^2 = 16$, with center at the origin and radius 4. This circle passes through $(0,4)$ and $(0,-4)$. As in choice (2), positive and negative values of $y$ are associated with the same value of $x$. Choice (4) is *not* a function.

The correct choice is **(3)**.

**12.** In the accompanying diagram of $\triangle ABC$ it is given that $m\angle A = 33$, $a = 12$, and $b = 15$. Let $x$ represent $m\angle B$:

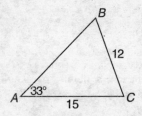

To find $m\angle B$ use the Law of Sines:

$$\frac{\sin 33}{12} = \frac{\sin x}{15}$$

Solve the equation for $\sin x$ and evaluate $\sin x$ with your calculator:

$$12 \sin x = 15 \sin 33$$

Find $x$ to the *nearest degree* using your calculator's $\sin^{-1}$ function:

$$\sin x = \frac{15 \sin 33}{12} \approx 0.6808$$

$$x = \sin^{-1}(0.6808) \approx 43°$$

The correct choice is (**2**).

**13.** In the accompanying diagram it is given that tangents $\overrightarrow{CA}$ and $\overrightarrow{CB}$ form an angle of 30°. When two tangents are drawn to a circle from the same external point, the tangent segments are congruent. Hence $\overline{CA} \cong \overline{CB}$, and $\triangle ACB$ is isosceles.

In an isosceles triangle the base angles are congruent; let $m\angle A = m\angle B = x$. The sum of the measures of the angles in any triangle is 180.

Solve for $x$:

$$x + x + 30 = 180$$
$$2x + 30 = 180$$
$$2x = 150$$
$$x = 75$$

The correct choice is (**3**).

**14.** Since sine functions pass through the origin and cosine functions have nonzero $y$-intercepts, the equation of the wave shown in the accompanying diagram is of the form $y = A \cos(Bx)$. $A$ determines the amplitude of the wave, and $B$ determines its period.

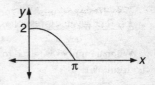

In the accompanying diagram, the amplitude of the curve is 2. Hence $A = 2$.

The given curve has completed one-quarter of a cycle when $x = \pi$, so its period is $4\pi$. Any function of the form $y = A \cos(Bx)$ completes one cycle when $Bx = 2\pi$. Solving $B(4\pi) = 2\pi$ yields $B = \frac{1}{2}$. Hence the equation of the given wave is $y = 2\cos(\frac{1}{2}x)$.

The correct choice is **(1)**.

**15.** Each time a bicycle tire rotates, the bicycle moves a distance equal to the circumference of the tire. It is given that the tires are 24 inches in diameter, which is equivalent to 2 feet. Hence the bicycle moves $\pi d = 2\pi$ feet during each tire rotation.

It is given that the tire rotates three times during each pedal rotation. Hence one pedal rotation moves the bicycle $3(2\pi) = 6\pi$ feet.

Since there are 5280 feet in a mile, the number of pedal rotations needed for the bicycle to travel at least 1 mile is $5280 \div 6\pi \approx 280.11$.

The correct choice is **(2)**.

**16.** In the accompanying diagram of $y = \cos x$, it can be seen that the curve has two kinds of symmetry.

• Vertical line symmetry (across the $y$-axis, for example).

• Point symmetry (with respect to $(\frac{\pi}{2}, 0)$, for example).

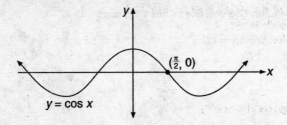

The correct choice is **(4)**.

**17.** The given expression is:

$$\left(\frac{3^0}{27^{\frac{2}{3}}}\right)^{-1}$$

When a number is raised to the 0 power, the result is 1.

$$\left(\frac{1}{27^{\frac{2}{3}}}\right)^{-1}$$

$$27^{\frac{2}{3}} = \left(\sqrt[3]{27}\right)^2 = 3^2 = 9$$

When a number is raised to the −1 power, the result is the reciprocal of that number.

$$\left(\frac{1}{9}\right)^{-1} = 9$$

The correct choice is (**2**).

**18.** The given function is:

$$h(x) = \sqrt{x^2 - 4x - 5}$$

The domain is the set of all values of $x$ such that:

$$x^2 - 4x - 5 \geq 0$$

To solve the quadratic inequality, first factor the trinomial into the product of two binomials:

$$(x + 1)(x - 5) \geq 0$$

This product is positive if both factors are positive or both are negative:

$x + 1 \geq 0$ and $x - 5 \geq 0$      $x + 1 \leq 0$ and $x - 5 \leq 0$

$x \geq -1$ and $x \geq 5$         $x \leq -1$ and $x \leq 5$

$x \geq 5$       OR      $x \leq -1$

The given function is defined when $x \geq 5$ or $x \leq -1$.

The correct choice is (**2**).

**19.** The given expression is:

$(-1 + i)^3$

Expand the binomial using $(a + b)^3 = a^3 + 3a^2b + 3ab^2 + b^3$:

$(-1)^3 + 3(-1)^2i + 3(-1)i^2 + i^3$

Simplify:

$-1 + 3i - 3i^2 + i^3$

$-1 + 3i - 3(-1) + (-i)$

$2 + 2i$

The correct choice is **(4)**.

**20.** It is given that revenue $R(x) = 35x$ and cost $C(x) = 20x + 500$. Then profit is represented by:

$$P(x) = R(x) - C(x)$$

$$= 35x - (20x + 500)$$

$$= 35x - 20x - 500$$

$$= 15x - 500$$

The correct choice is **(3)**.

## PART II

**21.** Two variables vary inversely if their product is a constant. The accompanying table shows values of $x$ and $y$ such that $xy = 12$. This is an example of inverse variation.

| $x$ | 1 | 2 | 3 | 4 | 6 | 12 |
|-----|----|---|---|---|---|----|
| $y$ | 12 | 6 | 4 | 3 | 2 | 1 |

**22.** The given equation is:

$$x^2 + 8x + 25 = 0$$

Solutions of a quadratic equation in the form $ax^2 + bx + c = 0$ are given by the quadratic formula:

$$x = \frac{-b \pm \sqrt{b^2 - 4ac}}{2a}$$

To use the formula to evaluate the roots of the given equation, substitute $a = 1$, $b = 8$, and $c = 25$:

$$= \frac{-8 \pm \sqrt{8^2 - 4 \cdot 1 \cdot 25}}{2 \cdot 1}$$

Simplify:

$$= \frac{-8 \pm \sqrt{64 - 100}}{2}$$

$$= \frac{-8 \pm \sqrt{-36}}{2}$$

$$= \frac{-8 \pm 6i}{2}$$

$$= -4 \pm 3i$$

In simplest form, the roots are **−4 + 3i** and **−4 − 3i**.

**23.** It is given that a ball has rolled 54° around a circle of radius 10 inches. The circumference of the circle is $C = 2\pi r = 2\pi(10) = 20\pi$ inches. The 54° arc represents $\dfrac{54}{360}$ of the circle, so its length is $\dfrac{54}{360}$ of the circumference:

$$\frac{54}{360}(20\pi) \approx 9.4248 \text{ inches}$$

To the *nearest hundredth* the ball has rolled **9.42 inches**.

**24.** When $w = 3$, the given expression is:

$$\frac{3}{h} = \frac{h}{3-h}$$

Cross-multiply and set the quadratic equal to 0:

$$h^2 = 9 - 3h$$

$$h^2 + 3h - 9 = 0$$

Solutions of a quadratic equation in the form $ax^2 + bx + c = 0$ are given by the quadratic formula:

$$h = \frac{-b \pm \sqrt{b^2 - 4ac}}{2a}$$

To use the formula to evaluate the roots, substitute $a = 1$, $b = 3$, and $c = -9$:

$$= \frac{-3 \pm \sqrt{3^2 - 4 \cdot 1 \cdot (-9)}}{2 \cdot 1}$$

Simplify:

$$= \frac{-3 \pm \sqrt{9 + 36}}{2}$$

$$= \frac{-3 \pm \sqrt{45}}{2}$$

Since $h$ represents the height of a rectangle, it must be positive. Use your calculator to evaluate the positive root:

$$= \frac{-3 + \sqrt{45}}{2} \approx 1.854$$

$h$ lies between the consecutive integers **1 and 2**.

**25.** It is given that, in $\triangle ABC$, $m\angle BAC = 62$, $AB = 12$, and $AC = 31$, as shown in the accompanying diagram.

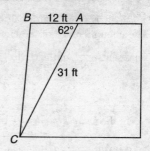

The area of a triangle is one-half the product of any two sides and the sine of the included angle:

Use your calculator to evaluate the area:

$$\text{Area } \triangle ABC = \frac{1}{2}(12)(31)\sin 62$$

$$\text{Area } \triangle ABC \approx 164.228$$

To the *nearest tenth* the area of $\triangle ABC$ is **164.2 square feet**.

**26.** It is given that student test scores are normally distributed with a mean of 80 and a standard deviation of 4. To determine the probability that a randomly selected score is between 74 and 82, first determine how far, in standard deviation units, each of these scores is from the mean.

- 74 is $\dfrac{74-80}{4} = -1.5$ standard deviations from (below) the mean.

- 82 is $\dfrac{82-80}{4} = 0.5$ standard deviations above the mean.

The accompanying figure shows that the percentage of scores between 1.5 standard deviations below the mean and 0.5 standard deviations above the mean is approximately:

$$9.2 + 15.0 + 19.1 + 19.1 = 62.4$$

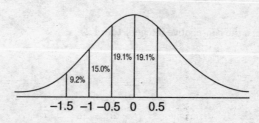

The probability is approximately **62.4%**.

# PART III

**27.** It is given that two roads intersect at a 40° angle, as shown in the accompanying diagram. It is also given that $PJ$ = 3.2 miles and $PM$ = 5.6 miles.

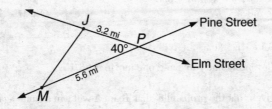

To find $JM$, use the Law of Cosines and your calculator:

$$JM^2 = PJ^2 + PM^2 - 2(PJ)(PM) \cos P$$
$$= (3.2)^2 + (5.6)^2 - 2(3.2)(5.6) \cos 40°$$
$$\approx 14.145$$
$$JM \approx \sqrt{14.145} \approx 3.761$$

To the *nearest tenth*, the distance is **3.8 miles**.

**28.** Let $M$ represent the cost of renting a movie and let $G$ represent the cost of renting a game. Express each of the given statements as an equation:

Jose rents two movies and three games for a total of $15.50:

$$2M + 3G = 15.50$$

Meg rents three movies and one game for a total of $12.05:

$$3M + G = 12.05$$

Solve the equations simultaneously by multiplying the second equation by –3 and then adding the two equations together:

$$2M + 3G = 15.50$$
$$\underline{-9M - 3G = -36.05}$$
$$-7M \quad\quad = -20.65$$

$$M = 2.95$$

Substitute the value of $M$ into one of the original equations and find the value of $G$:

$$3(2.95) + G = 12.05$$

$$G = 3.20$$

Find the cost of renting one game and one movie:

$$G + M = 3.20 + 2.95$$

$$= 6.15$$

The cost of renting one game and one movie is **$6.15**.

**29.** It is given that the probability that Team A will win a game against Team B is $\frac{1}{3}$.

To find the probability that Team A will win exactly 3 of 5 games, evaluate the probability formula $_nC_x p^x(1-p)^{n-x}$, where $n = 5$, $x = 3$, and $p = \frac{1}{3}$.

$$P(x = 3) = {_5}C_3(\tfrac{1}{3})^3(\tfrac{2}{3})^2 = \frac{40}{243}$$

To find the probability that Team A will win exactly 4 games, evaluate the probability formula using $x = 4$:

$$P(x = 4) = {_5}C_4(\tfrac{1}{3})^4(\tfrac{2}{3})^1 = \frac{10}{243}$$

To find the probability that Team A will win all 5 games, evaluate the probability formula using $x = 5$:

$$P(x = 5) = {_5}C_5(\tfrac{1}{3})^5(\tfrac{2}{3})^0 = \frac{1}{243}$$

The probability that Team A will win at least 3 of 5 games is the sum of these probabilities:

$$P(x \geq 3) = P(x = 3) + P(x = 4) + P(x = 5) = \frac{40}{243} + \frac{10}{243} + \frac{1}{243} = \frac{51}{243}$$

The probability is $\frac{51}{243}$.

**30.** The given formula is:

$$V = C(1 - r)^t$$

To find the age of the car, substitute the original cost of the car, $C = 15,000$, the rate of depreciation, $r = 0.30$, and the current value of the car, $V = 3000$:

$$3000 = 15000(1 - 0.30)^t$$

Solve the equation for $t$:

$$0.2 = 0.7^t$$

$$\log(0.2) = t \log(0.7)$$

$$t = \frac{\log(0.2)}{\log(0.7)}$$

Evaluate using your calculator:

$$t \approx 4.512$$

To the *nearest tenth*, the age of the car is **4.5** years.

**31.** The given function is:

$$h(t) = -16t^2 + 64t + 4$$

To determine when the height of the ball is 52 feet, let $h = 52$:

$$52 = -16t^2 + 64t + 4$$

To solve the quadratic equation, first rewrite it as a trinomial that equals 0:

$$16t^2 - 64t + 48 = 0$$

The terms of the trinomial have 16 as a common factor, and the remaining trinomial can be factored into the product of two binomials:

$$16(t^2 - 4t + 3) = 0$$

$$16(t - 1)(t - 3) = 0$$

If the product of two factors is zero, then at least one of them must equal zero. Set each factor equal to 0 and solve for $t$:

$$t - 1 = 0 \quad \text{or} \quad t - 3 = 0$$

$$t = 1 \quad \text{or} \quad t = 3$$

Since the height of the baseball is exactly 52 feet at time $t = 1$ (on the way up) and again at time $t = 3$ (on the way down), its height is greater than 52 feet between these two times.

The interval of time is $1 \leq t \leq 3$.

**32.** To graph the equation $2y = 2x^2 - 4$, first simplify by dividing both sides of the equation by 2: $y = x^2 - 2$.

List the integer values in the interval $-3 \leq x \leq 3$ and find the resulting $y$-coordinates, as shown in the table.

| $x$ | $y = x^2 - 2$ | $y$ | $(x,y)$ |
|-----|---------------|-----|---------|
| $-3$ | $(-3)^2 - 2 = 9 - 2 = 7$ | 7 | $(-3,7)$ |
| $-2$ | $(-2)^2 - 2 = 4 - 2 = 2$ | 2 | $(-2,2)$ |
| $-1$ | $(-1)^2 - 2 = 1 - 2 = -1$ | $-1$ | $(-1,-1)$ |
| $0$ | $0^2 - 2 = 0 - 2 = -2$ | $-2$ | $(0,-2)$ |
| $1$ | $1^2 - 2 = 1 - 2 = -1$ | $-1$ | $(1,-1)$ |
| $2$ | $2^2 - 2 = 4 - 2 = 2$ | 2 | $(2,2)$ |
| $3$ | $3^2 - 2 = 9 - 2 = 7$ | 7 | $(3,7)$ |

Using graph paper, plot the points listed in the $(x,y)$ column of the table and connect them with a smooth curve in the shape of a parabola with its vertex at $(0,-2)$. Label the graph $a$, as shown in the accompanying diagram.

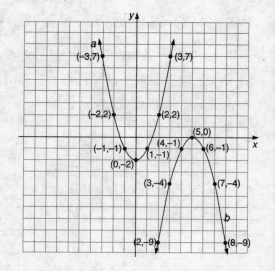

To find the image of $a$ under $T_{5,-2} \circ r_{x\text{-}axis}$, first find the image of each point after a reflection in the $x$-axis using the transformation $(x,y) \rightarrow (x,-y)$, as shown in the table.

Now translate that image $T_{5,-2}$ using the transformation $(x,y) \rightarrow (x + 5, y - 2)$ as shown in the table.

Graph the image by plotting the points in the last column and connecting them with a smooth curve. Label this image $b$, as shown in the diagram.

| $a$ | $r_{x\text{-}axis}$ | $T_{5,-2}$ |
|-----|---------------------|------------|
| $(-3,7)$ | $(-3,-7)$ | $(2,-9)$ |
| $(-2,2)$ | $(-2,-2)$ | $(3,-4)$ |
| $(-1,-1)$ | $(-1,1)$ | $(4,-1)$ |
| $(0,-2)$ | $(0,2)$ | $(5,0)$ |
| $(1,-1)$ | $(1,1)$ | $(6,-1)$ |
| $(2,2)$ | $(2,-2)$ | $(7,-4)$ |
| $(3,7)$ | $(3,-7)$ | $(8,-9)$ |

# PART IV

**33.** Given: Parallelogram $ABCD$
            Diagonals $\overline{AC}$ and $\overline{BD}$
            intersect at $E$

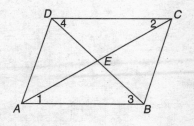

Prove: Diagonals $\overline{AC}$ and $\overline{BD}$ bisect
each other.

Plan: Show that $\triangle ABE \cong \triangle CDE$
by ASA $\cong$ ASA and then show that
$\overline{AE} \cong \overline{CE}$ and $\overline{BE} \cong \overline{DE}$ by CPCTC.

**PROOF**

| Statement | Reason |
|---|---|
| 1. $ABCD$ is a parallelogram | 1. Given |
| 2. $\overline{AB} \cong \overline{CD}$       (side) | 2. Opposite sides of a parallelogram are congruent. |
| 3. $\overline{AB} \parallel \overline{CD}$ | 3. Opposite sides of a parallelogram are parallel. |
| 4. $\angle 1 \cong \angle 2$      (angle)<br>    $\angle 3 \cong \angle 4$      (angle) | 4. When parallel lines are cut by a transversal, the alternate interior angles are congruent. |
| 5. $\triangle ABE \cong \triangle CDE$ | 5. ASA $\cong$ ASA |
| 6. $\overline{AE} \cong \overline{CE}$ and $\overline{BE} \cong \overline{DE}$ | 6. CPCTC |
| 7. Diagonals $\overline{AC}$ and $\overline{BD}$ bisect each other. | 7. The bisector of a segment divides it into two congruent segments. |

**34.** To construct a scatterplot of the given data, plot the test scores for each person as the point (Test $x$, Test $y$) as shown in the accompanying diagram.

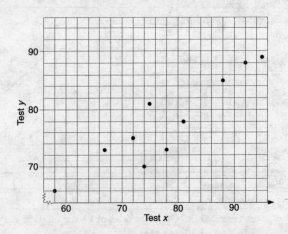

Enter the given data into your calculator with Test $x$ as $x$ and Test $y$ as $y$ and use the calculator to find the equation of the line of best fit.

The equation is $y = 0.62x + 29.18$.

Use your calculator to find the correlation coefficient. The correlation is $r = 0.92$.

To predict the score on test $y$ for a student who scored 87 on test $x$, substitute $x = 87$ into the equation and evaluate $y$: $y = 0.62(87) + 29.18 = 83.12$.

The model predicts that the student's score, to the *nearest integer*, will be **83**.

| Topic | Question Numbers | Number of Points | Your Points |
|---|---|---|---|
| 1. Properties of Real Numbers | — | — | |
| 2. Sequences | — | — | |
| 3. Complex Numbers | 19 | 2 | |
| 4. Inequalities, Absolute Value | 18 | 2 | |
| 5. Algebraic Expressions, Fractions | 2, 6, 20 | 2 + 2 + 2 = 6 | |
| 6. Exponents (zero, negative, rational, scientific notation) | 17 | 2 | |
| 7. Radical Expressions | — | — | |
| 8. Quadratic Equations (factors, formula, discriminant) | 1, 22, 24, 31 | 2 + 2 + 2 + 4 = 10 | |
| 9. Systems of Equations | 28 | 4 | |
| 10. Functions (graphs, domain, range, roots) | 11 | 2 | |
| 11. Inverse Functions, Composition | 7, 9 | 2 + 2 = 4 | |
| 12. Linear Functions | — | — | |
| 13. Parabolas (max/min, axis of symmetry) | 32a | 2 | |
| 14. Hyperbolas (including inverse variation) | 21 | 2 | |
| 15. Ellipses | — | — | |
| 16. Exponents and Logarithms | 3, 8, 30 | 2 + 2 + 4 = 8 | |
| 17. Trig. (circular functions, unit circle, radians) | 4, 5, 14, 25 | 2 + 2 + 2 + 2 = 8 | |
| 18. Trig. Equations and Identities | — | — | |
| 19. Solving Triangles (sin/cos laws, Pythag. thm., ambig. case) | 12, 27 | 2 + 4 = 6 | |
| 20. Coordinate Geom. (slope, distance, midpoint, circle) | — | — | |
| 21. Transformations, Symmetry | 10, 16, 32b | 2 + 2 + 2 = 6 | |
| 22. Circle Geometry | 13, 23 | 2 + 2 = 4 | |
| 23. Congruence, Similarity, Proportions | — | — | |
| 24. Probability (including Bernoulli events) | 29 | 4 | |
| 25. Normal Curve | 26 | 2 | |
| 26. Statistics (center, spread, summation notation) | — | — | |
| 27. Correlation, Modeling, Prediction | 34 | 6 | |
| 28. Algebraic Proofs, Word Problems | 15 | 2 | |
| 29. Geometric Proofs | 33 | 6 | |
| 30. Coordinate Geometry Proofs | — | — | |

**Total Raw Score =** _____

## HOW TO CONVERT YOUR RAW SCORE
## TO YOUR MATH B REGENTS EXAMINATION SCORE

Below is the conversion chart that must be used to determine your final score on the January 2002 Regents Examination in Math B. To find your final exam score, locate in the column labeled "Raw Score" the total number of points you scored. Then locate in the adjacent column to the right the scaled score that corresponds to your raw score. The scaled score is your final Math B Regents Examination score.

### Regents Examination in Math B—January 2002
### Chart for Converting Total Test Raw Scores to
### Final Examination Scores (Scaled Scores)

| Raw Score | Scaled Score | Raw Score | Scaled Score | Raw Score | Scaled Score |
|---|---|---|---|---|---|
| 88 | 100 | 58 | 81 | 28 | 47 |
| 87 | 99 | 57 | 80 | 27 | 46 |
| 86 | 99 | 56 | 79 | 26 | 44 |
| 85 | 99 | 55 | 78 | 25 | 43 |
| 84 | 98 | 54 | 77 | 24 | 42 |
| 83 | 98 | 53 | 76 | 23 | 40 |
| 82 | 97 | 52 | 75 | 22 | 39 |
| 81 | 97 | 51 | 74 | 21 | 37 |
| 80 | 96 | 50 | 74 | 20 | 35 |
| 79 | 96 | 49 | 73 | 19 | 34 |
| 78 | 95 | 48 | 71 | 18 | 32 |
| 77 | 94 | 47 | 70 | 17 | 31 |
| 76 | 94 | 46 | 69 | 16 | 29 |
| 75 | 93 | 45 | 68 | 15 | 27 |
| 74 | 93 | 44 | 67 | 14 | 26 |
| 73 | 92 | 43 | 66 | 13 | 24 |
| 72 | 91 | 42 | 65 | 12 | 22 |
| 71 | 91 | 41 | 64 | 11 | 21 |
| 70 | 90 | 40 | 63 | 10 | 19 |
| 69 | 89 | 39 | 62 | 9 | 17 |
| 68 | 89 | 38 | 60 | 8 | 15 |
| 67 | 88 | 37 | 59 | 7 | 13 |
| 66 | 87 | 36 | 58 | 6 | 12 |
| 65 | 87 | 35 | 57 | 5 | 10 |
| 64 | 86 | 34 | 55 | 4 | 8 |
| 63 | 85 | 33 | 54 | 3 | 6 |
| 62 | 84 | 32 | 53 | 2 | 4 |
| 61 | 83 | 31 | 51 | 1 | 2 |
| 60 | 83 | 30 | 50 | 0 | 0 |
| 59 | 82 | 29 | 49 | | |

# Examination
# June 2002
## Math B

**FORMULAS**

### Area of Triangle

$$K = \frac{1}{2}ab \sin C$$

### Functions of the Sum of Two Angles

$$\sin (A + B) = \sin A \cos B + \cos A \sin B$$
$$\cos (A + B) = \cos A \cos B - \sin A \sin B$$

### Functions of the Difference of Two Angles

$$\sin (A - B) = \sin A \cos B - \cos A \sin B$$
$$\cos (A - B) = \cos A \cos B + \sin A \sin B$$

### Law of Sines

$$\frac{a}{\sin A} = \frac{b}{\sin B} = \frac{c}{\sin C}$$

### Law of Cosines

$$a^2 = b^2 + c^2 - 2bc \cos A$$

### Functions of the Double Angle

$$\sin 2A = 2 \sin A \cos A$$
$$\cos 2A = \cos^2 A - \sin^2 A$$
$$\cos 2A = 2 \cos^2 A - 1$$
$$\cos 2A = 1 - 2 \sin^2 A$$

## Functions of the Half Angle

$$\sin \frac{1}{2}A = \pm\sqrt{\frac{1 - \cos A}{2}}$$

$$\cos \frac{1}{2}A = \pm\sqrt{\frac{1 + \cos A}{2}}$$

## Normal Curve
## Standard Deviation

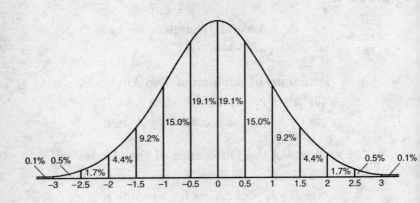

# PART I

Answer all questions in this part. Each correct answer will receive 2 credits. No partial credit will be allowed. Record your answers in the spaces provided. [40]

1 What is the value of $\sum_{m=2}^{5} (m^2 - 1)$?

(1) 58            (3) 53

(2) 54            (4) 50          1 _____

2 For all values of $x$ for which the expression is defined, $\dfrac{2x + x^2}{x^2 + 5x + 6}$ is equivalent to

(1) $\dfrac{1}{x+3}$            (3) $\dfrac{1}{x+2}$

(2) $\dfrac{x}{x+3}$            (4) $\dfrac{x}{x+2}$          2 _____

3 In the accompanying diagram, the length of $\overset{\frown}{ABC}$ is $\dfrac{3\pi}{2}$ radians.

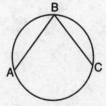

(Not drawn to scale)

What is m$\angle ABC$?

(1) 36            (3) 53

(2) 45            (4) 72          3 _____

4 In the accompanying diagram of $\triangle ABC$, $\overline{AB} \cong \overline{AC}$, $\overline{BD} = \frac{1}{3}\overline{BA}$, and $\overline{CE} = \frac{1}{3}\overline{CA}$.

Triangle $EBC$ can be proved congruent to triangle $DCB$ by

(1) SAS $\cong$ SAS      (3) SSS $\cong$ SSS

(2) ASA $\cong$ ASA      (4) HL $\cong$ HL     4 _____

5 The path of a rocket is represented by the equation $y = \sqrt{25-x^2}$ . The path of a missile designed to interest the path of the rocket is represented by the equation $x = \frac{3}{2}\sqrt{y}$ . The value of $x$ at the point of intersection is 3. What is the corresponding value of $y$?

(1) –2      (3) –4

(2) 2      (4) 4     5 _____

6 On a standardized test, the distribution of scores is normal, the mean of the scores is 75, and the standard deviation is 5.8. If a student scored 83, the student's score ranks

(1) below the 75th percentile
(2) between the 75th percentile and the 84th percentile
(3) between the 84th percentile and the 97th percentile
(4) above the 97th percentile

6 _____

7 Which statement is true for all real number values of $x$?

(1) $|x - 1| > 0$          (3) $\sqrt{x^2} = x$

(2) $|x - 1| > (x - 1)$      (4) $\sqrt{x^2} = |x|$

7 _____

8 If $x$ is a positive integer, $4x^{\frac{1}{2}}$ is equivalent to

(1) $\dfrac{2}{x}$          (3) $4\sqrt{x}$

(2) $2x$            (4) $4\dfrac{1}{x}$

8 _____

9 What is the equation of a parabola that goes through points $(0,1)$, $(-1,6)$.

(1) $y = x^2 + 1$       (3) $y = x^2 - 3x + 1$
(2) $y = 2x^2 + 1$     (4) $y = 2x^2 - 3x + 1$

9 _____

10 If $f(x) = 2x^2 + 4$ and $g(x) = x - 3$, which number satisfies $f(x) = (f \circ g)(x)$?

(1) $\dfrac{3}{2}$        (3) 5

(2) $\dfrac{3}{4}$        (4) 4      10 _____

11 A linear regression equation of best fit between a student's attendance and the degree of success in school is $h = 0.5x + 68.5$. The correlation coefficient, $r$, for these data would be

(1) $0 < r < 1$        (3) $r = 0$
(2) $-1 < r < 0$        (4) $r = -1$      11 _____

12 What is the solution set of the equation $\dfrac{x}{x - 4} - \dfrac{1}{x + 3} = \dfrac{28}{x^2 - x - 12}$?

(1) { }        (3) {–6}
(2) {4,–6}        (4) {4}      12 _____

13 Which equation represents a function?

(1) $4y^2 = 36 - 9x^2$        (3) $x^2 + y^2 = 4$
(2) $y = x^2 - 3x - 4$        (4) $x = y^2 - 6x + 8$      13 _____

14 What is the solution set of the equation $x = 2\sqrt{2x - 3}$?

(1) { }        (3) {6}
(2) {2}        (4) {2,6}      14 _____

15 What is the sum of $\sqrt{-2}$ and $\sqrt{-18}$ ?

(1) $5i\sqrt{2}$             (3) $2i\sqrt{5}$

(2) $4i\sqrt{2}$             (4) $6i$           15 _____

16 Which diagram represents a one-to-one function?

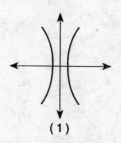

(1)

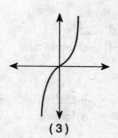

(3)

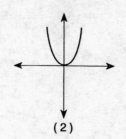

(2)

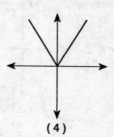

(4)

16 _____

17 Point $P'$ is the image of point $P(-3,4)$ after a translation defined by $T_{(7,-1)}$. Which other transformation on $P$ would also produce $P'$?

(1) $r_{y\,=\,-x}$             (3) $R_{90°}$

(2) $r_{y\text{-axis}}$             (4) $R_{-90°}$        17 _____

18 Which transformation does *not* preserve orientation?

    (1) translation
    (2) dilation
    (3) reflection in the $y$-axis
    (4) rotation          18 ____

19 The roots of the equation $2x^2 - x = 4$ are

    (1) real and irrational
    (2) real, rational, and equal
    (3) real, rational, and unequal
    (4) imaginary          19 ____

20 Which graph represents the inverse of
   $f(x) = \{(0,1),(1,4),(2,3)\}$?

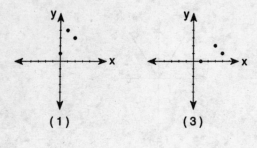

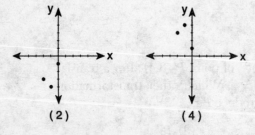

20 ____

## PART II

Answer all questions in this part. Each correct answer will receive 2 credits. Clearly indicate the necessary steps, including appropriate formula substitutions, diagrams, graphs, charts, etc. For all questions in this part, a correct numerical answer with no work shown will receive only 1 credit. [12]

21 On a nationwide examination, the Adams School had a mean score of 875 and a standard deviation of 12. The Boswell School had a mean score of 855 and a standard deviation of 20. In which school was there greater consistency in the scores? Explain how you arrived at your answer.

22 Is $\frac{1}{2} \sin 2x$ the same expression as $\sin x$? Justify your answer.

23 After studying a couple's family history, a doctor determines that the probability of any child born to this couple having a gene for disease $X$ is 1 out of 4. If the couple has three children, what is the probability that *exactly* two of the children have the gene for disease $X$?

24 Growth of a certain strain of bacteria is modeled by the equation $G = A(2.7)^{0.584t}$, where:

$G$ = final number of bacteria
$A$ = initial number of bacteria
$t$ = time (in hours)

In approximately how many hours will 4 bacteria first increase to 2,500 bacteria? Round your answer to the *nearest hour.*

25 The equation $W = 120I - 12I^2$ represents the power ($W$), in watts, of a 120-volt circuit having a resistance of 12 ohms when a current ($I$) is flowing through the circuit. What is the maximum power, in watts, that can be delivered in this circuit?

26  Island Rent-a-Car charges a car rental fee of $40 plus
    $5 per hour or fraction of an hour. Wayne's Wheels
    charges a car rental fee of $25 plus $7.50 per hour or
    fraction of an hour. Under what conditions does it
    cost *less* to rent from Island Rent-a-Car? [The use of
    the accompanying grid is optional.]

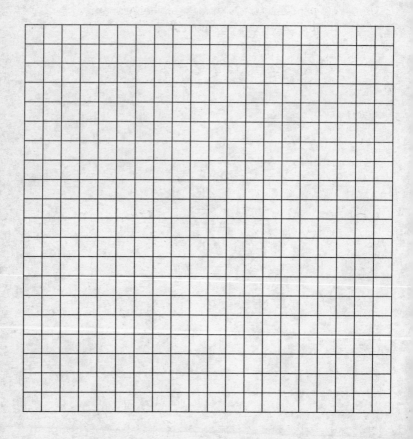

## PART III

**Answer all questions in this part. Each correct answer will receive 4 credits. Clearly indicate the necessary steps, including appropriate formula substitutions, diagrams, graphs, charts, etc. For all questions in this part, a correct numerical answer with no work shown will receive only 1 credit.** [24]

27 An electronics company produces a headphone set that can be adjusted to accommodate different-sized heads. Research into the distance between the top of people's heads and the top of their ears produced the following data, in inches:

4.5, 4.8, 6.2, 5.5, 5.6, 5.4, 5.8, 6.0, 5.8, 6.2, 4.6, 5.0, 5.4, 5.8

The company decides to design their headphones to accommodate three standard deviations from the mean. Find, to the *nearest tenth*, the mean, the standard deviation, and the range of distances that must be accommodated.

28  A pelican flying in the air over water drops a crab
    from a height of 30 feet. The distance the crab is
    from the water as it falls can be represented by the
    function $h(t) = -16t^2 + 30$, where $t$ is time, in sec-
    onds. To catch the crab as it falls, a gull flies along a
    path represented by the function $g(t) = -8t + 15$.
    Can the gull catch the crab before the crab hits the
    water? Justify your answer. [The use of the accom-
    panying grid is optional.]

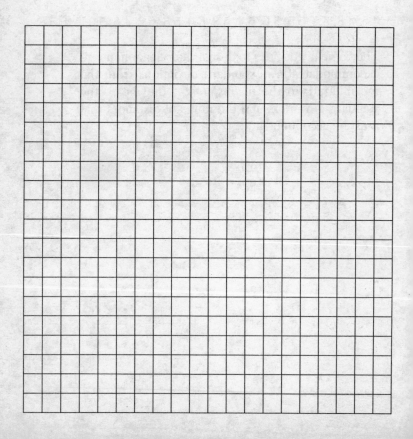

29 Complete the partial proof below for the accompanying diagram by providing reasons for steps 3, 6, 8, and 9.

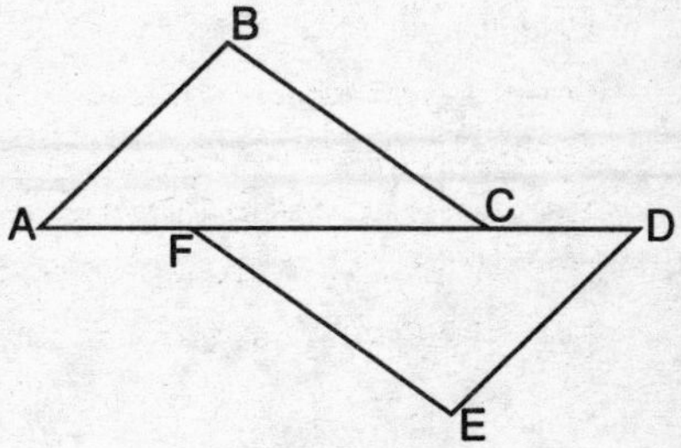

Given: $\overline{AFCD}$
$\overline{AB} \perp \overline{BC}$
$\overline{DE} \perp \overline{EF}$
$\overline{BC} \parallel \overline{FE}$
$\overline{AB} \cong \overline{DE}$

Prove: $\overline{AC} \cong \overline{FD}$

| Statements | Reasons |
|---|---|
| 1 $\overline{AFCD}$ | 1 Given |
| 2 $\overline{AB} \perp \overline{BC}, \overline{DE} \perp \overline{EF}$ | 2 Given |
| 3 $\angle B$ and $\angle E$ are right angles. | 3 _____ |
| 4 $\angle B \cong \angle E$ | 4 All right angles are congruent. |
| 5 $\overline{BC} \parallel \overline{FE}$ | 5 Given |
| 6 $\angle BCA \cong \angle EFD$ | 6 _____ |
| 7 $\overline{AB} \cong \overline{DE}$ | 7 Given |
| 8 $\triangle ABC \cong \triangle DEF$ | 8 _____ |
| 9 $\overline{AC} \cong \overline{FD}$ | 9 _____ |

30 Solve for $x$:   $\log_4 (x^2 + 3x) - \log_4 (x + 5) = 1$

31 A ship at sea heads directly toward a cliff on the shoreline. The accompanying diagram shows the top of the cliff, $D$, sighted from two locations, $A$ and $B$, separated by distance $S$. If m$\angle DAC = 30$, m$\angle DBC = 45$, and $S = 30$ feet, what is the height of the cliff, to the *nearest foot*?

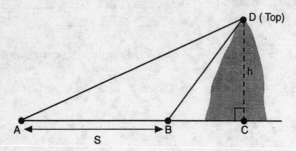

32 Kieran is traveling from city $A$ to city $B$. As the accompanying map indicates, Kieran could drive directly from $A$ to $B$ along County Route 21 at an average speed of 55 miles per hour or travel on the interstates, 45 miles along I-85 and 20 miles along I-64. The two interstates intersect at an angle of 150° at $C$ and have a speed limit of 65 miles per hour. How much time will Kieran save by traveling along the interstates at an average speed of 65 miles per hour?

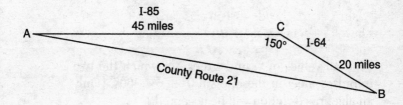

## PART IV

**Answer all questions in this part. Each correct answer will receive 6 credits. Clearly indicate the necessary steps, including appropriate formula substitutions, diagrams, graphs, charts, etc. For all questions in this part, a correct numerical answer with no work shown will receive only 1 credit.**   [12]

33 On a monitor, the graphs of two impulses are recorded on the same screen, where $0° \leq x < 360°$. The impulses are given by the following equations:

$$y = 2 \sin^2 x$$
$$y = 1 - \sin x$$

Find all values of $x$, in degrees, for which the two impulses meet in the interval $0° \leq x < 360°$. [Only an algebraic solution will be accepted.]

34 The table below, created in 1996, shows a history of transit fares from 1955 to 1995. On the accompanying grid, construct a scatter plot where the independent variable is years. State the exponential regression equation with the coefficient and base rounded to the *nearest thousandth*. Using this equation, determine the prediction that should have been made for the year 1998, to the *nearest cent*.

| Year | 55 | 60 | 65 | 70 | 75 | 80 | 85 | 90 | 95 |
|---|---|---|---|---|---|---|---|---|---|
| Fare ($) | 0.10 | 0.15 | 0.20 | 0.30 | 0.40 | 0.60 | 0.80 | 1.15 | 1.50 |

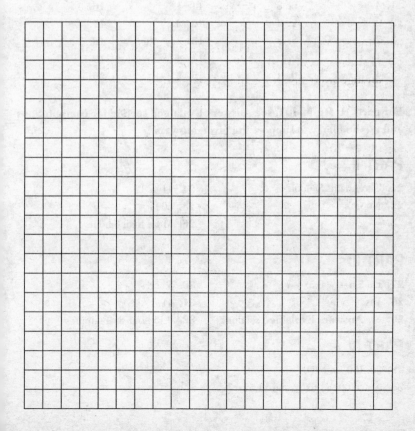

# Answers
# June 2002
## Math B

## Answer Key

## PART I

| | | | |
|---|---|---|---|
| 1. 4 | 6. 3 | 11. 1 | 16. 3 |
| 2. 2 | 7. 4 | 12. 3 | 17. 4 |
| 3. 2 | 8. 3 | 13. 2 | 18. 3 |
| 4. 1 | 9. 4 | 14. 4 | 19. 1 |
| 5. 4 | 10. 1 | 15. 2 | 20. 3 |

**For parts II, III, and IV, see Answers Explained section for computations and methodology to support the given answers.**

## PART II

**21.** The Adams School

**22.** No

**23.** $\dfrac{9}{64}$

**24.** 11 or 12

**25.** 300

**26.** More than 6 hours

## PART III

**27.** $\bar{x} = 5.5$, $\sigma = 0.5$, 4–7

**28.** Yes

**29.** *See* **Answers Explained** section.

**30.** 5 and –4

**31.** 41

**32.** 0.15 hour or 9 minutes

## PART IV

**33.** 30, 150, and 270

**34.** $y = (0.002)(1.070)^x$ and $1.52

# Answers Explained

## PART I

**1.** The given expression $\sum\limits_{m=2}^{5} (m^2 - 1)$ represents the sum of the terms $m^2 - 1$ for integer values of $m$ from 2 to 5. To evaluate the summation, substitute each of the values $m = 2, 3, 4,$ and 5 into the expression and add the results:

$$\sum_{m=2}^{5} (m^2 - 1) = (2^2 - 1) + (3^2 - 1) + (4^2 - 1) + (5^2 - 1)$$

$$= (4 - 1) + (9 - 1) + (16 - 1) + (25 - 1)$$

$$= 3 + 8 + 15 + 24$$

$$= 50$$

The correct choice is (**4**).

**2.** The given expression is:

$$\frac{2x + x^2}{x^2 + 5x + 6}$$

To reduce this fraction, first factor the numerator and denominator. In the numerator, $x$ is a common factor of both terms. In the denominator, factor the trinomial into the product of two binomials.

$$\frac{x(2 + x)}{(x + 2)(x + 3)}$$

Since $x + 2 = 2 + x$, reduce the fraction by canceling these factors in the numerator and denominator:

$$\frac{x(2 + x)}{(x + 2)(x + 3)} = \frac{x}{x + 3}$$

The correct choice is (**2**).

**3.** In the accompanying diagram it is given that the length of $\widehat{ABC}$ is $\frac{3\pi}{2}$ radians. The entire circle measures $2\pi$ radians, hence the length of minor arc $\widehat{AC}$ is $2\pi - \frac{3\pi}{2} = \frac{\pi}{2}$ radians. In a circle the measure of an inscribed angle is half the measure of the intercepted arc. $\frac{\pi}{2}$ radians is equivalent to 90°, hence

$$m\angle ABC = \tfrac{1}{2}m\widehat{AC} = \tfrac{1}{2}(90) = 45$$

The correct choice is (**2**).

**(Not drawn to scale)**

**4.** It is given in the accompanying diagram that $\overline{AB} \cong \overline{AC}$. If two sides of a triangle are congruent, the angles opposite them are congruent; hence $\angle DBC \cong \angle ECB$. (Angle)

It is also given that $\overline{BD}$ and $\overline{CE}$ are each $\frac{1}{3}$ of the congruent sides $\overline{AB}$ and $\overline{AC}$. Hence $\overline{BD} \cong \overline{CE}$. (Side)

$\triangle EBC$ and $\triangle DCB$ have $\overline{BC}$ in common. (Side)

Hence $\triangle EBC$ and $\triangle DCB$ can be proved congruent by $SAS \cong SAS$.

The correct choice is (**1**).

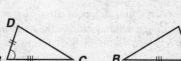

**5.** The path of the rocket is represented by $y = \sqrt{25 - x^2}$. To find the value of $y$ that corresponds to $x = 3$, substitute:

$$y = \sqrt{25 - x^2} = \sqrt{25 - 3^2} = \sqrt{25 - 9} = \sqrt{16} = 4$$

The correct choice is (**4**).

**6.** It is given that test scores are normally distributed with a mean of 75 and a standard deviation of 5.8. To determine the percentile rank of a student score of 83, first determine how far, in standard deviation units, the score is from the mean.

83 is $\dfrac{83 - 75}{5.8} \approx 1.38$ standard deviations above the mean.

The accompanying figure shows that the percentage of scores below 1.00 standard deviation above the mean is approximately $50 + 19.1 + 15.0 = 84.1\%$. It also shows that $84.1 + 9.2 = 93.3\%$ of the scores are below 1.50 standard deviations above the mean.

Since the student score of 83 is between 1.00 and 1.50 standard deviations above the mean, it ranks between the 84th and 93rd percentiles.

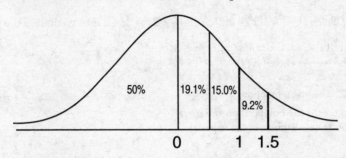

The correct choice is **(3)**.

**7.** Consider the choices offered:

- Choice (1): if $x = 1$, $|1 - 1| = 0$; hence $|x - 1| > 0$ is not true for all real values of $x$.
- Choice (2): if $x = 1$, $|1 - 1| = 1 - 1$; hence $|x - 1| > (x - 1)$ is not true for all real values of $x$.
- Choice (3): if $x = -1$, $\sqrt{(-1)^2} = 1$; hence $\sqrt{x^2} = x$ is not true for all real values of $x$.
- Choice (4): since $\sqrt{x^2}$ and $|x|$ are always nonnegative, $\sqrt{x^2} = |x|$ *is* true for all real values of $x$.

The correct choice is **(4)**.

**8.** Since $x^{\frac{1}{2}} = \sqrt{x}$, $4x^{\frac{1}{2}} = 4\sqrt{x}$.

The correct choice is **(3)**.

**9.** It is given that the parabola passes through the point $(2,3)$; hence the values $x = 2$ and $y = 3$ must satisfy the equation. To check each equation substitute $x = 2$ and $y = 3$:

- Choice (1): $3 \neq 2^2 + 1 = 5$, so $y = x^2 + 1$ does not go through the point $(2,3)$.
- Choice (2): $3 \neq 2(2^2) + 1 = 9$, so $y = 2x^2 + 1$ does not go through the point $(2,3)$.
- Choice (3): $3 \neq 2^2 - 3(2) + 1 = -1$, so $y = x^2 - 3x + 1$ does not go through the point $(2,3)$.
- Choice (4): $3 = 2(2^2) - 3(2) + 1$, so $y = 2x^2 - 3x + 1$ *does* go through the point $(2,3)$.

The correct choice is **(4)**.

**10.** The given functions are:           $f(x) = 2x^2 + 4$       $g(x) = x - 3$

To create the composition $(f \circ g)(x)$, substitute $(x - 3)$ for $x$ in the expression for $f(x)$:

$$(f \circ g)(x) = 2(x - 3)^2 + 4$$

Write the equation $f(x) = (f \circ g)(x)$:     $2x^2 + 4 = 2(x - 3)^2 + 4$

Simplify, and solve for $x$:            $2x^2 + 4 = 2(x^2 - 6x + 9) + 4$

$$4 = -12x + 22$$

$$12x = 18$$

$$x = \frac{3}{2}$$

The correct choice is **(1)**.

**11.** The given linear regression equation, $h = 0.5x + 68.5$, has a positive slope of $0.5$. For any set of data the correlation coefficient, $r$, and the slope of the line of best fit have the same sign. Hence, $r$ must also be positive.

The correct choice is **(1)**.

**12.** The given equation is:

$$\frac{x}{x-4} - \frac{1}{x+3} = \frac{28}{x^2 - x - 12}$$

To find the common denominator, factor the trinomial into the product of two binomials:

$$\frac{x}{x-4} - \frac{1}{x+3} = \frac{28}{(x-4)(x+3)}$$

To solve an equation containing fractions, multiply both sides of the equation by the common denominator, here $(x-4)(x+3)$:

$$\frac{x}{x-4}(x-4)(x+3) - \frac{1}{x+3}(x-4)(x+3) = \frac{28}{(x-4)(x+3)}(x-4)(x+3)$$

In each term cancel the common factors:

$$\frac{x}{\cancel{x-4}}(\cancel{x-4})(x+3) - \frac{1}{\cancel{x+3}}(x-4)(\cancel{x+3}) = \frac{28}{\cancel{(x-4)(x+3)}}\cancel{(x-4)(x+3)}$$

Simplify:

$$x(x+3) - (x-4) = 28$$

$$x^2 + 3x - x + 4 = 28$$

Rewrite the equation as a trinomial that equals 0:

$$x^2 + 2x - 24 = 0$$

Factor the trinomial into the product of two binomials:

$$(x+6)(x-4) = 0$$

If the product of two quantities is 0, then at least one of the quantities must equal 0. Solve each equation for $x$:

$$x + 6 = 0 \qquad x - 4 = 0$$
$$x = -6 \qquad \cancel{x = 4}$$

Note that the denominator of the first fraction, $x - 4$, would equal 0 if $x = 4$; hence 4 is an extraneous root of the equation. The only root is $x = -6$.

The correct choice is **(3)**.

**13.** In a function each value of $x$ is associated with only one value of $y$. In each of the equations in choices (1), (3), and (4), $y$ appears only in terms containing $y^2$. Hence, positive and negative values of $y$ are associated with the same value of $x$. The equations in choices (1), (3), and (4) do *not* represent functions.

Choice (2) is the parabola $y = x^2 - 3x - 4$. For each value of $x$ there is only one value of $y$ that determines a point on the parabola. Choice (2) *does* represent a function.

The correct choice is **(3)**.

**14.** The given equation is:
$$x = 2\sqrt{2x - 3}$$

To solve an equation involving a radical, square both sides and then simplify:
$$x^2 = 4(2x - 3)$$
$$= 8x - 12$$

To solve the quadratic equation, rewrite it in standard form with 0 on one side and all other terms on the other side.
$$x^2 - 8x + 12 = 0$$

Factor the quadratic trinomial as the product of two binomials:
$$(x - 6)(x - 2) = 0$$

If the product of two quantities is 0, then at least one of them is 0. Solve the two equations for $x$:
$$x - 6 = 0 \text{ or } x - 2 = 0$$
$$x = 6 \quad \text{or} \quad x = 2$$

Since squaring both sides of an equation can introduce extraneous roots, check both solutions in the original equation:

If $x = 6$:         If $x = 2$:

$6 = 2\sqrt{2(6) - 3}$?    $2 = 2\sqrt{2(2) - 3}$?

$6 = 2\sqrt{9}$ ✔        $2 = 2\sqrt{1}$ ✔

Both $x = 6$ and $x = 2$ are roots of the equation.

The correct choice is **(4)**.

**15.** The given sum is:

$$\sqrt{-2} + \sqrt{-18}$$

Rewrite each term as an imaginary number in simplest radical form:

$$\sqrt{(-1)(2)} + \sqrt{(-1)(2)(9)}$$

Add the like terms:

$$i\sqrt{2} + 3i\sqrt{2}$$

$$4i\sqrt{2}$$

The correct choice is (**2**).

**16.** In a one-to-one function each value of $x$ is associated with only one value of $y$. For choice (3) any horizontal line such as the one shown in the accompanying diagram intersects the graph only once, identifying the unique value of $x$ associated with that value of $y$. Hence this diagram represents a one-to-one function.

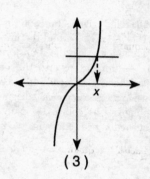

(3)

The correct choice is (**3**).

**17.** The translation $T_{(7,-1)}$ is defined by $(x,y) \rightarrow (x + 7, x - 1)$. After this translation the image of $P(-3,4)$ is $P'(4,3)$.

The image of $P(-3,4)$ is also $P'(4,3)$ after the transformation $(x,y) \rightarrow (y,-x)$, which represents the rotation $R_{-90°}$.

The correct choice is (**4**).

**18.** Translations, dilations, and rotations all preserve the orientation of a figure, but reflections do not.

The correct choice is (**3**).

**19.** The given quadratic equation, $2x^2 - x = 4$, can be rewritten as $2x^2 - x - 4 = 0$. This equation is in standard form $ax^2 + bx + c = 0$ with $a = 2$, $b = -1$ and $c = -4$.

The nature of the roots can be determined by evaluating the discriminant:

$$b^2 - 4ac = (-1)^2 - 4(2)(-4) = 33$$

The discriminant is the expression under the square root in the quadratic formula. Because 33 is positive, the roots are real. Because 33 is not a perfect square, the roots are irrational.

The correct choice is **(1)**.

**20.** The given function is $f = \{(0,1),(1,4),(2,3)\}$ To find the inverse of any function, use the transformation $(x,y) \rightarrow (y,x)$ to reverse the ordered pairs. Thus $f^{-1} = \{(1,0),(4,1),(3,2)\}$. As shown in the accompanying diagram, the graph given as choice (3) plots these points and hence represents the inverse of $f$.

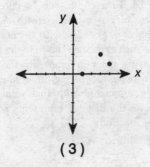

**( 3 )**

The correct choice is **(3)**.

## PART II

**21.** It is given that the standard deviations of exam scores were 12 at the Adams School and 20 at the Boswell School. Standard deviations measure the spread of the scores, reflecting how far scores scatter from the mean. Since the standard deviation was smaller at the Adams School, the variability among scores was less there.

There was greater consistency in the scores at the **Adams School**.

**22.** Suppose the given expressions are the same:

$$\frac{1}{2}\sin 2x = \sin x$$

Rewrite the equation using the identity for $\frac{1}{2}\sin 2x$ and simplify:

$$\frac{1}{2}(2\sin x \cos x) = \sin x$$

$$\sin x \cos x = \sin x$$

Rewrite the equation so that it equals 0:

$$\sin x \cos x - \sin x = 0$$

The two terms on the left side of the equation have $\sin x$ as a common factor:

$$\sin x(\cos x - 1) = 0$$

If the product of two quantities is 0, at least one of them is 0.

$$\sin x = 0 \text{ or } \cos x - 1 = 0$$
$$\sin x = 0 \text{ or } \cos x = 1$$

The original equation is true only when $\sin x = 0$ or $\cos x = 1$, not for all values of $x$.

No, $\frac{1}{2}\sin 2x$ **is *not* the same expression as** $\sin x$.

**23.** It is given that the probability that the child has the gene is $\frac{1}{4}$. To find the probability that exactly 2 of the 3 children have the gene, evaluate the probability formula $_nC_x p^x(1-p)^{n-x}$, where $n = 3$, $x = 2$, and $p = \frac{1}{4}$:

$$P(x = 3) = {}_3C_2\left(\frac{1}{4}\right)^2\left(\frac{3}{4}\right)^1 = \frac{9}{64}$$

The probability is $\frac{9}{64}$.

**24.** The given formula is:

$$G = A(2.7)^{0.584t}$$

To find the number of hours it will take until the number of bacteria increases from the original 4 to 2500, substitute $A = 4$ and $G = 2500$:

$$2500 = 4(2.7)^{0.584t}$$

Solve the equation for $t$:

$$625 = 2.7^{0.584t}$$

$$\log(625) = 0.584t \log(2.7)$$

$$t = \frac{\log(625)}{0.584 \ \log(2.7)}$$

Evaluate using your calculator:

$$t \approx 11.098$$

Note that after 11 hours there are not quite 2500 bacteria, hence the population first increases to 2500 after the 12th hour. However, this question was vaguely worded in asking for an *approximate* number of hours (11.098 is very close to 11), so the Regents granted full credit for an answer of 11 or 12.

To the *nearest hour*, the number of hours is **11** or **12**.

**25.** The power of the circuit (in watts) is given by $W = 120I - 12I^2$ , which is a quadratic equation. The graph of a quadratic equation of the form $y = ax^2 + bx + c$ is a parabola. The maximum value of a parabola is at its vertex (turning point). The vertex lies on the axis of symmetry, whose equation is $x = -\dfrac{b}{2a}$.

To find the value of $I$ at the turning point substitute $a = -12$ and $b = 120$:

$I = -\dfrac{b}{2a} = \dfrac{120}{2(-12)} = 5$. Therefore, the circuit attains its maximum power when $I = 5$.

To find the maximum power substitute $I = 5$ in the equation for power:

$$W = 120I - 12I^2 = 120(5) - 12(5)^2 = 300 \text{ watts}$$

The maximum power is **300 watts**.

**26.** It is given that Island Rent-a-Car charges $40 plus $5 per hour, and Wayne's Wheels charges $25 plus $7.50 per hour.

Let $h$ = the number of hours the car is rented. Then the costs of the two plans can be represented by:     $I = 40 + 5h$ and $W = 25 + 7.5h$

Island costs less when:     $40 + 5h < 25 + 7.5h$

Solve for $h$:     $-2.5h < -15$

    $h > 6$

It costs *less* to rent from Island for **more than 6 hours**.

# PART III

**27.** Use your calculator to find that the mean and standard deviation of the given data, to the *nearest tenth*, are $\bar{x} = 5.5$ inches and $\sigma = 0.5$ inches.

- The distance three standard deviations below the mean is $5.5 - 3(0.5) = 4$ inches.
- The distance three standard deviations above the mean is $5.5 + 3(0.5) = 7$ inches.

The range of distances that must be accommodated is **4–7 inches**.

**28.** The given functions representing the heights of the crab and the gull at time $t$ are:

$$h(t) = -16t^2 + 30$$
$$g(t) = -8t + 15$$

The gull can catch the crab only if it is at the same height at the same time, so set the expressions equal and solve the resulting quadratic trinomial by factoring it into the product of two binomials:

$$-8t + 15 = -16t^2 + 30$$
$$16t^2 - 8t - 15 = 0$$
$$(4t + 3)(4t - 5) = 0$$

If the product of two quantities is 0, at least one of them is 0. Solve each equation for $t$:

$$4t + 3 = 0 \quad \text{or} \quad 4t - 5 = 0$$
$$t = -\frac{4}{3} \quad \text{or} \quad t = \frac{5}{4}$$

Only $t = \frac{5}{4}$ is a possibility. But is the gull still above the level of the water? To find out, substitute $t = \frac{5}{4}$ in the equation for the height of the gull:

$$g\left(\frac{5}{4}\right) = -8\left(\frac{5}{4}\right) + 15$$

$$g\left(\frac{5}{4}\right) = 5 \text{ feet}$$

**Yes**, the gull can catch the falling crab.

**29.** The reasons for the four steps are:

- Step 3: Perpendicular line segments form right angles.
- Step 6: When parallel lines are cut by a transversal, the alternate interior angles formed are congruent.
- Step 8: AAS $\cong$ AAS
- Step 9: Corresponding parts of congruent triangles are congruent.

**30.** The given equation is:
$$\log_4(x^2 + 3x) - \log_4(x + 5) = 1$$

$$\log A - \log B = \log \frac{A}{B}$$

$$\log_4 \frac{x^2 + 3x}{x + 5} = 1$$

Rewrite the logarithm equation as an equivalent exponential equation:

$$\frac{x^2 + 3x}{x + 5} = 4^1$$

Rewrite the resulting quadratic equation as a trinomial equal to 0:

$$x^2 + 3x = 4(x + 5)$$
$$x^2 + 3x = 4x + 20$$
$$x^2 - x - 20 = 0$$

Factor the quadratic trinomial as the product of two binomials:

$$(x + 4)(x - 5) = 0$$

If the product of two quantities is 0, at least one of them is 0. Solve each equation for $x$:

$$x + 4 = 0 \quad \text{or} \quad x - 5 = 0$$
$$x = -4 \quad \text{or} \quad x = 5$$

The values of $x$ are **-4** and **5**.

**31.** In the accompanying diagram it is given that m∠DAC = 30, m∠DBC = 45, and S = 30 feet.

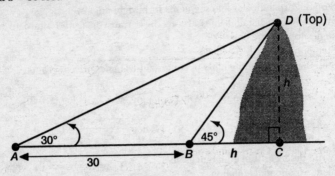

Since ∠C is a right angle and m∠DBC = 45, △BCD is an isosceles right triangle; hence BC = DC = h.

Since m∠DAC = 30, △ACD is a 30-60-90 triangle with sides in the ratio 1:2:$\sqrt{3}$ .

$$\frac{AC}{DC} = \frac{\sqrt{3}}{1}$$

$$\frac{30 + h}{h} = \frac{\sqrt{3}}{1}$$

In any proportion, the product of the means equals the product of the extremes, so cross-multiply and solve for h:

$$h\sqrt{3} = 30 + h$$

$$h\sqrt{3} - h = 30$$

$$h(\sqrt{3} - 1) = 30$$

$$h = \frac{30}{\sqrt{3} - 1}$$

Use your calculator to find the value of h:

$$h \approx 40.98$$

To the *nearest foot* the height of the cliff is **41 feet**.

**32.** In the accompanying diagram it is given that $AC = 45$, $BC = 20$, and $m\angle C = 150°$.

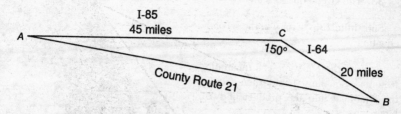

Use the Law of Cosines and your calculator to estimate distance $AB$:

$$(AB)^2 = (AC)^2 + (BC)^2 - 2(AC)(BC)\cos C$$

$$= (45)^2 + (20)^2 - 2(45)(20)\cos 150°$$

$$= 2025 + 400 - 1800(-\frac{\sqrt{3}}{2})$$

$$\approx 3983.846$$

$$AB \approx 63.118 \text{ miles}$$

If Kieran travels directly from $A$ to $B$ at 55 miles per hour, it will take approximately $63.118 \div 55 = 1.15$ hours.

If Kieran travels from $A$ to $C$ and then on to $B$, a total distance of 65 miles, at 65 miles per hour, it will take 1 hour.

Kieran will save approximately **0.15 hour** or **9 minutes** by taking the interstates.

# PART IV

**33.** The given equations are:

$$y = 2\sin^2 x \qquad y = 1 - \sin x$$

Substitute $2\sin^2 x$ for $y$ in the second equation:

$$2\sin^2 x = 1 - \sin x$$

Rewrite the equation as a quadratic trinomial equaling 0 and then factor the trinomial into the product of two binomials:

$$2\sin^2 x + \sin x - 1 = 0$$

$$(2\sin x - 1)(\sin x + 1) = 0$$

If the product of two quantities is 0, at least one of them is 0. Solve each equation for $\sin x$:

$$2\sin x - 1 = 0 \quad \text{or} \quad \sin x + 1 = 0$$
$$\sin x = \tfrac{1}{2} \qquad \text{or} \qquad \sin x = -1$$

Determine all angles between $0°$ and $360°$ whose sines equal $\tfrac{1}{2}$ or $-1$:

$$x = 30°, 150°, 270°$$

$$x = 30°, 150°, 270°$$

The values of $x$ are **30°**, **150°**, and **270°**.

**34.** To construct a scatterplot of the given data, plot the point (year, fare) for each of the years given as shown in the accompanying diagram.

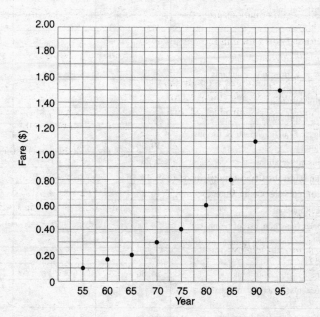

Enter the given data into your calculator with the year as $x$ and the fare as $y$ and use the calculator to find the exponential regression equation.

The equation, with coefficient and base rounded to the *nearest thousandth*, is $y = 0.002(1.070)^x$.

To predict the fare for the year 1998, substitute $x = 98$ into the equation and evaluate $y$:

$$y = 0.002(1.070)^{98} \approx 1.516$$

The model predicts that the 1998 fare, to the *nearest cent*, was **$1.52**.

| Topic | Question Numbers | Number of Points | Your Points |
|---|---|---|---|
| 1. Properties of Real Numbers | — | — | |
| 2. Sequences | — | — | |
| 3. Complex Numbers | 15 | **2** | |
| 4. Inequalities, Absolute Value | 7 | **2** | |
| 5. Algebraic Expressions, Fractions | 2, 12 | 2 + 2 = **4** | |
| 6. Exponents (zero, negative, rational, scientific notation) | — | — | |
| 7. Radical Expressions | 8, 14 | 2 + 2 = **4** | |
| 8. Quadratic Equations (factors, formula, discriminant) | 19, 30 | 2 + 2 = **4** | |
| 9. Systems of Equations | 5, 28 | 2 + 4 = **6** | |
| 10. Functions (graphs, domain, range, roots) | 9, 13, 16 | 2 + 2 + 2 = **6** | |
| 11. Inverse Functions, Composition | 10, 20 | 2 + 2 = **4** | |
| 12. Linear Functions | 26 | **2** | |
| 13. Parabolas (max/min, axis of symmetry) | 25 | **2** | |
| 14. Hyperbolas (including inverse variation) | — | — | |
| 15. Ellipses | — | — | |
| 16. Exponents and Logarithms | 24, 30 | 2 + 2 = **4** | |
| 17. Trig. (circular functions, unit circle, radians) | — | — | |
| 18. Trig. Equations and Identities | 22, 33 | 2 + 4 = **6** | |
| 19. Solving Triangles (sin/cos laws, Pythag. Thm., Ambig. Case) | 31, 32 | 4 + 4 = **8** | |
| 20. Coordinate Geom. (slope, distance, midpoint, circle) | — | — | |
| 21. Transformations, Symmetry | 17, 18 | 2 + 2 = **4** | |
| 22. Circle Geometry | 3 | **2** | |
| 23. Congruence, Similarity, Proportions | 4 | **2** | |
| 24. Probability (including Bernoulli events) | 23 | **2** | |
| 25. Normal Curve | 6 | **2** | |
| 26. Statistics (center, spread, summation notation) | 1, 21, 27 | 2 + 2 + 4 = **8** | |
| 27. Correlation, Modeling, Prediction | 11, 34 | 2 + 6 = **8** | |
| 28. Algebraic Proofs, Word Problems | — | — | |
| 29. Geometric Proofs | 29 | **4** | |
| 30. Coordinate Geometry Proofs | — | — | |

**Total Raw Score =** _____

## HOW TO CONVERT YOUR RAW SCORE
## TO YOUR MATH B REGENTS EXAMINATION SCORE

Below is the conversion chart that must be used to determine your final score on the June 2002 Regents Examination in Math B. To find your final exam score, locate in the column labeled "Raw Score" the total number of points you scored. Then locate in the adjacent column to the right the scaled score that corresponds to your raw score. The scaled score is your final Math B Regents Examination score.

**Regents Examination in Math B—June 2002**
**Chart for Converting Total Test Raw Scores to**
**Final Examination Scores (Scaled Scores)**

| Raw Score | Scaled Score | Raw Score | Scaled Score | Raw Score | Scaled Score |
|---|---|---|---|---|---|
| 88 | 100 | 58 | 80 | 28 | 54 |
| 87 | 99 | 57 | 79 | 27 | 53 |
| 86 | 98 | 56 | 78 | 26 | 52 |
| 85 | 97 | 55 | 78 | 25 | 50 |
| 84 | 96 | 54 | 77 | 24 | 49 |
| 83 | 96 | 53 | 77 | 23 | 48 |
| 82 | 95 | 52 | 76 | 22 | 46 |
| 81 | 94 | 51 | 75 | 21 | 45 |
| 80 | 93 | 50 | 75 | 20 | 43 |
| 79 | 93 | 49 | 74 | 19 | 41 |
| 78 | 92 | 48 | 73 | 18 | 40 |
| 77 | 91 | 47 | 72 | 17 | 38 |
| 76 | 90 | 46 | 72 | 16 | 36 |
| 75 | 90 | 45 | 71 | 15 | 34 |
| 74 | 89 | 44 | 70 | 14 | 33 |
| 73 | 88 | 43 | 69 | 13 | 31 |
| 72 | 88 | 42 | 69 | 12 | 29 |
| 71 | 87 | 41 | 68 | 11 | 27 |
| 70 | 87 | 40 | 67 | 10 | 24 |
| 69 | 86 | 39 | 66 | 9 | 22 |
| 68 | 85 | 38 | 65 | 8 | 20 |
| 67 | 85 | 37 | 64 | 7 | 18 |
| 66 | 84 | 36 | 63 | 6 | 15 |
| 65 | 84 | 35 | 62 | 5 | 13 |
| 64 | 83 | 34 | 61 | 4 | 11 |
| 63 | 82 | 33 | 60 | 3 | 8 |
| 62 | 82 | 32 | 59 | 2 | 5 |
| 61 | 81 | 31 | 58 | 1 | 3 |
| 60 | 81 | 30 | 57 | 0 | 0 |
| 59 | 80 | 29 | 56 | | |

# Examination August 2002

## Math B

## FORMULAS

### Area of Triangle

$$K = \frac{1}{2}ab\sin C$$

### Function of the Sum of Two Angles

$$\sin (A + B) = \sin A \cos B + \cos A \sin B$$
$$\cos (A + B) = \cos A \cos B - \sin A \sin B$$

### Function of the Difference of Two Angles

$$\sin (A - B) = \sin A \cos B - \cos A \sin B$$
$$\cos (A - B) = \cos A \cos B + \sin A \sin B$$

### Law of Sines

$$\frac{a}{\sin A} = \frac{b}{\sin B} = \frac{c}{\sin C}$$

### Law of Cosines

$$a^2 = b^2 + c^2 - 2bc \cos A$$

### Functions of the Double Angle

$$\sin 2A = 2 \sin A \cos A$$
$$\cos 2A = \cos^2 A - \sin^2 A$$
$$\cos 2A = 2 \cos^2 A - 1$$
$$\cos 2A = 1 - 2 \sin^2 A$$

## Functions of the Half Angle

$$\sin \frac{1}{2}A = \pm\sqrt{\frac{1 - \cos A}{2}}$$

$$\cos \frac{1}{2}A = \pm\sqrt{\frac{1 + \cos A}{2}}$$

## Normal Curve
## Standard Deviation

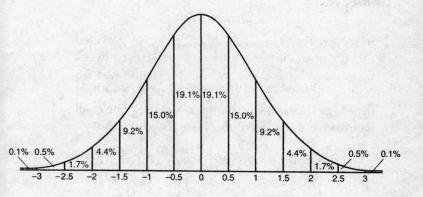

# PART I

Answer all questions in this part. Each correct answer will receive 2 credits. No partial credit will be allowed. Record your answers in the spaces provided. [40]

1 Which fraction represents the probability of obtaining *exactly* eight heads in ten tosses of a fair coin?

(1) $\dfrac{45}{1,024}$          (3) $\dfrac{90}{1,024}$

(2) $\dfrac{64}{1,024}$          (4) $\dfrac{180}{1,024}$      1 _____

2 In a New York City high school, a survey revealed the mean amount of cola consumed each week was 12 bottles and the standard deviation was 2.8 bottles. Assuming the survey represents a normal distribution, how many bottles of cola per week will approximately 68.2% of the students drink?

(1) 6.4 to 12          (3) 9.2 to 14.8

(2) 6.4 to 17.6        (4) 12 to 20.4      2 _____

3 What is the solution of the inequality $|x + 3| \leq 5$?

(1) $-8 \leq x \leq 2$          (3) $x \leq -8$ or $x \geq 2$

(2) $-2 \leq x \leq 8$          (4) $x \leq -2$ or $x \geq 8$      3 _____

4 What is the domain of $f(x) = 2^x$?

(1) all integers          (3) $x \geq 0$

(2) all real numbers      (4) $x \leq 0$      4 _____

5 A function is defined by the equation $y = 5x - 5$. Which equation defines the inverse of this function?

(1) $y = \dfrac{1}{5x - 5}$          (3) $x = \dfrac{1}{5y - 5}$

(2) $y = 5x + 5$          (4) $x = 5y - 5$      5 _____

6 An architect is designing a building to include an arch in the shape of a semi-ellipse (half an ellipse), such that the width of the arch is 20 feet and the height of the arch is 8 feet, as shown in the accompanying diagram.

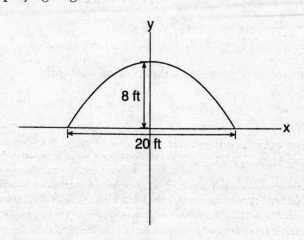

Which equation models this arch?

(1) $\dfrac{x^2}{100}+\dfrac{y^2}{64}=1$      (3) $\dfrac{x^2}{64}+\dfrac{y^2}{100}=1$

(2) $\dfrac{x^2}{400}+\dfrac{y^2}{64}=1$      (4) $\dfrac{x^2}{64}+\dfrac{y^2}{400}=1$    6 _____

7 To balance a seesaw, the distance, in feet, a person is from the fulcrum is inversely proportional to the person's weight, in pounds. Bill, who weighs 150 pounds, is sitting 4 feet away from the fulcrum. If Dan weighs 120 pounds, how far from the fulcrum should he sit to balance the seesaw?

(1) 4.5 ft      (3) 3 ft
(2) 3.5 ft      (4) 5 ft    7 _____

8 What is the *last* term in the expansion of $(x + 2y)^5$?

    (1) $y^5$                 (3) $10y^5$

    (2) $2y^5$              (4) $32y^5$          8 _____

9 In the equation $\log_x 4 + \log_x 9 = 2$, $x$ is equal to

    (1) $\sqrt{13}$             (3) 6.5

    (2) 6                (4) 18           9 _____

10 Which expression represents the sum of $\dfrac{1}{\sqrt{3}}$ and $\dfrac{1}{\sqrt{2}}$?

    (1) $\dfrac{2\sqrt{3} + 3\sqrt{2}}{6}$         (3) $\dfrac{\sqrt{3} + \sqrt{2}}{3}$

    (2) $\dfrac{2}{\sqrt{5}}$              (4) $\dfrac{\sqrt{3} + \sqrt{2}}{2}$     10 _____

11 Which equation has imaginary roots?

    (1) $x^2 - 1 = 0$         (3) $x^2 + x + 1 = 0$

    (2) $x^2 - 2 = 0$         (4) $x^2 - x - 1 = 0$    11 _____

12 If $\log k = c \log v + \log p$, $k$ equals

    (1) $v^c p$              (3) $v^c + p$

    (2) $(vp)^c$           (4) $cv + p$       12 _____

13 If $_nC_r$ represents the number of combinations of $n$ items taken $r$ at a time, what is the value of $\displaystyle\sum_{r=1}^{3} {}_4C_r$?

    (1) 24              (3) 6

    (2) 14             (4) 4         13 _____

14 In the accompanying diagram of $\triangle ABC$, m$\angle A = 30$, m$\angle C = 50$, and $AC = 13$.

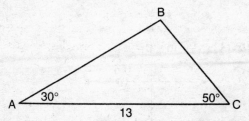

What is the length of side $\overline{AB}$ to the *nearest tenth*?

(1) 6.6                    (3) 11.5
(2) 10.1                   (4) 12.0                    14 _____

15 Expressed in simplest form, $i^{16} + i^6 - 2i^5 + i^{13}$ is equivalent to

(1) 1                     (3) $i$
(2) −1                    (4) $-i$                     15 _____

16 If point $(a,b)$ lies on the graph $y = \mathrm{f}(x)$, the graph $y = \mathrm{f}^{-1}(x)$ must contain point

(1) $(b,a)$               (3) $(0,b)$
(2) $(a,0)$               (4) $(-a,-b)$                16 _____

17 If the sum of the roots of $x^2 + 3x - 5 = 0$ is added to the product of its roots, the result is

(1) 15                    (3) −2
(2) −15                   (4) −8                       17 _____

18 The expression $\dfrac{3^{\frac{1}{3}}}{3^{-\frac{2}{3}}}$ is equivalent to

(1) 1                     (3) 3

(2) $\sqrt{3}$            (4) $\dfrac{1}{\sqrt[3]{3}}$  18 _____

19 The accompanying graph represents the figure $\text{I}$.

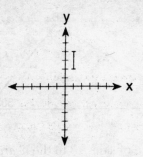

Which graph represents $\text{I}$ after a transformation defined by $r_{y=x} \circ R_{90°}$?

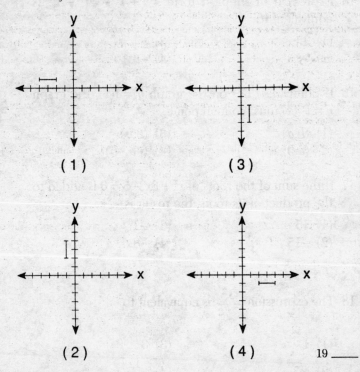

(1)

(3)

(2)

(4)

19 _____

20 Which expression is equivalent to the complex

fraction $\dfrac{\dfrac{x}{x+2}}{1-\dfrac{x}{x+2}}$ ?

(1) $\dfrac{2}{x}$          (3) $\dfrac{2x}{x+2}$

(2) $\dfrac{x}{2}$          (4) $\dfrac{2x}{x^2+4}$        20 _____

## PART II

Answer all questions in this part. Each correct answer will receive 2 credits. Clearly indicate the necessary steps, including appropriate formula substitutions, diagrams, graphs, charts, etc. For all questions in this part, a correct numerical answer with no work shown will receive only 1 credit.   [12]

21  A used car was purchased in July 1999 for $11,900. If the car depreciates 13% of its value each year, what is the value of the car, to the *nearest hundred dollars*, in July 2002?

22  The amount of time that a teenager plays video games in any given week is normally distributed. If a teenager plays video games an average of 15 hours per week, with a standard deviation of 3 hours, what is the probability of a teenager playing video games between 15 and 18 hours a week?

23 An art student wants to make a string collage by connecting six equally spaced points on the circumference of a circle to its center with string. What would be the radian measure of the angle between two adjacent pieces of string, in simplest form?

24 The Franklins inherited $3,500, which they want to invest for their child's future college expenses. If they invest it at 8.25% with interest compounded monthly, determine the value of the account, in dollars, after 5 years. Use the formula $A = P\left(1+\dfrac{r}{n}\right)^{nt}$, where $A$ = value of the investment after $t$ years, $P$ = principal invested, $r$ = annual interest rate, and $n$ = number of times compounded per year.

25 A toy truck is located within a circular play area. Alex and Dominic are sitting on opposite endpoints of a chord that contains the truck. Alex is 4 feet from the truck, and Dominic is 3 feet from the truck. Meira and Tamara are sitting on opposite endpoints of another chord containing the truck. Meira is 8 feet from the truck. How many feet, to the *nearest tenth of a foot*, is Tamara from the truck? Draw a diagram to support your answer.

26 Two sides of a triangular-shaped pool measure 16 feet and 21 feet, and the included angle measures 58°. What is the area, to the *nearest tenth of a square foot*, of a nylon cover that would exactly cover the surface of the pool?

# PART III

**Answer all questions in this part. Each correct answer will receive 4 credits. Clearly indicate the necessary steps, including appropriate formula substitutions, diagrams, graphs, charts, etc. For all questions in this part, a correct numerical answer with no work shown will receive only 1 credit.** [24]

27 The cost $(C)$ of selling $x$ calculators in a store is modeled by the equation $C = \dfrac{3,200,000}{x} + 60,000$.

The store profit $(P)$ for these sales is modeled by the equation $P = 500x$. What is the minimum number of calculators that have to be sold for profit to be greater than cost?

28 Two tow trucks try to pull a car out of a ditch. One tow truck applies a force of 1,500 pounds while the other truck applies a force of 2,000 pounds. The resultant force is 3,000 pounds. Find the angle between the two applied forces, rounded to the *nearest degree*.

29 A rock is thrown vertically from the ground with a velocity of 24 meters per second, and it reaches a height of $2 + 24t - 4.9t^2$ after $t$ seconds. How many seconds after the rock is thrown will it reach maximum height, and what is the maximum height the rock will reach, in meters? How many seconds after the rock is thrown will it hit the ground? Round your answers to the *nearest hundredth*. [Only an algebraic or graphic solution will be accepted.]

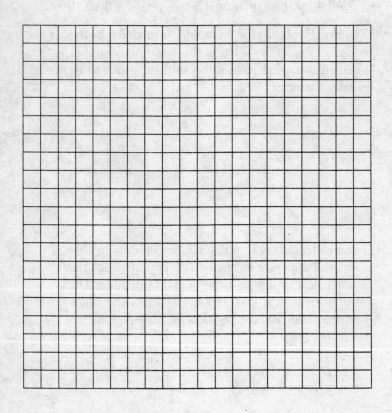

30 In the accompanying diagram, △*ABC* is *not* isosceles. Prove that if altitude $\overline{BD}$ were drawn, it would
*not* bisect $\overline{AC}$.

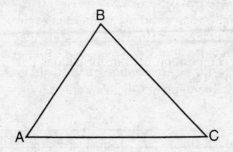

31  Graph and label the following equations, $a$ and $b$, on the accompanying set of coordinate axes.

$$a: y = x^2$$
$$b: y = -(x-4)^2 + 3$$

Describe the composition of transformations performed on $a$ to get $b$.

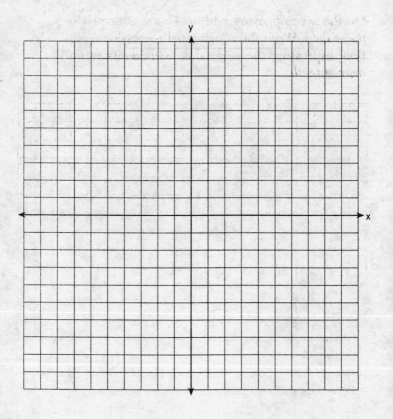

32 The breaking strength, $y$, in tons, of steel cable with diameter $d$, in inches, is given in the table below.

| $d$ (in) | 0.50 | 0.75 | 1.00 | 1.25 | 1.50 | 1.75 |
|---|---|---|---|---|---|---|
| $y$ (tons) | 9.85 | 21.80 | 38.30 | 59.20 | 84.40 | 114.00 |

On the accompanying grid, make a scatter plot of these data. Write the exponential regression equation, expressing the regression coefficients to the *nearest tenth*.

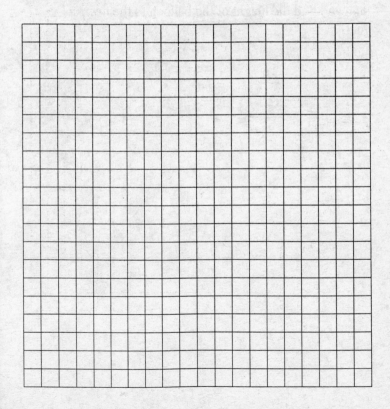

## PART IV

Answer all questions in this part. Each correct answer will receive 6 credits. Clearly indicate the necessary steps, including appropriate formula substitutions, diagrams, graphs,
charts, etc. For all questions in this part, a correct numerical
answer with no work shown will receive only 1 credit.   [12]

33  Carmen and Jamal are standing 5,280 feet apart on
    a straight, horizontal road. They observe a hot-air
    balloon between them directly above the road. The
    angle of elevation from Carmen is 60° and from
    Jamal is 75°. Draw a diagram to illustrate this situa-
    tion and find the height of the balloon to the *nearest
    foot*.

34 Electrical circuits can be connected in series, one after another, or in parallel circuits that branch off a main line. If circuits are hooked up in parallel, the reciprocal of the total resistance in the series is found by adding the reciprocals of each resistance, as shown in the accompanying diagram.

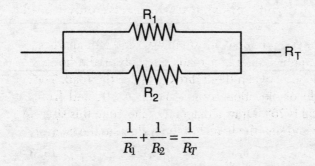

$$\frac{1}{R_1} + \frac{1}{R_2} = \frac{1}{R_T}$$

If $R_1 = x$, $R_2 = x + 3$, and the total resistance, $R_T$, is 2.25 ohms, find the positive value of $R_1$ to the *nearest tenth of an ohm*.

# Answers
# August 2002
## Math B

## Answer Key

### PART I

| | | | |
|---|---|---|---|
| 1. 1 | 6. 1 | 11. 3 | 16. 1 |
| 2. 3 | 7. 4 | 12. 1 | 17. 4 |
| 3. 1 | 8. 4 | 13. 2 | 18. 3 |
| 4. 2 | 9. 2 | 14. 2 | 19. 3 |
| 5. 4 | 10. 1 | 15. 4 | 20. 2 |

For Parts II, III, and IV, see Answers Explained section for computations and methodology to support the given answers.

### PART II

21. 7,800

22. 0.341 or 34.1%

23. $\dfrac{\pi}{3}$

24. 5,279.61

25. 1.5

26. 142.5

### PART III

27. 161

28. 63

29. time of max height = 2.45
maximum height = 31.39
time hitting the ground = 4.98

30. See **Answers Explained** section.

31. See **Answers Explained** section.

32. $y = 4.8(6.8)^x$

### PART IV

33. 6,246

34. 3.5

# Answers Explained

## PART I

**1.** When a fair coin is tossed, the probability of getting a head is $\frac{1}{2}$. To find the probability of obtaining exactly 8 heads in 10 tosses, evaluate the probability formula $_nC_x p^x(1-p)^{n-x}$, where $n = 10$, $x = 8$, and $p = \frac{1}{2}$:

$$P(x = 8) = {_{10}}C_8\left(\frac{1}{2}\right)^8\left(1 - \frac{1}{2}\right)^2\left(\frac{1}{2}\right)^2 = \frac{45}{1,024}$$

The correct choice is **(1)**.

**2.** It is given that the amount of cola consumed each week was normally distributed with a mean of 12 bottles per week and a standard deviation of 2.8 bottles. In a normal distribution approximately 68.2% of the individuals lie within ±1 standard deviation of the mean. Therefore, this interval, $12 \pm 1(2.8) = 9.2$ to $14.8$, represents the number of bottles of cola per week that 68.2% of the students drink.

The correct choice is **(3)**.

**3.** The given inequality is:                                  $|x + 3| \le 5$
Rewrite the inequality without absolute value:          $-5 \le x + 3 \le 5$
Solve the inequality:                                        $-8 \le x \le 2$

The correct choice is **(1)**.

**4.** The given equation, $f(x) = 2^x$, is an exponential function. The domain is all real numbers.

The correct choice is **(2)**.

**5.** The given function is:                                $y = 5x - 5$
To find an equation of the inverse of this function,
interchange $x$ and $y$:                                      $x = 5y - 5$

The correct choice is **(4)**.

**6.** It is given that an arch is in the shape of a semi-ellipse with height 8 feet and width 20 feet, as shown in the accompanying diagram.

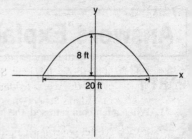

The equation of an ellipse with semi-major axis $a$ and semi-minor axis $b$ is:

$$\frac{x^2}{a^2} + \frac{y^2}{b^2} = 1$$

The major axis of the given ellipse is 20, the width of the arch, and the semi-minor axis is 8, the height of the arch. To find the equation that models this arch, substitute $a = 10$ and $b = 8$:

$$\frac{x^2}{10^2} + \frac{y^2}{8^2} = 1$$

$$\frac{x^2}{100} + \frac{y^2}{64} = 1$$

The correct choice is **(1)**.

**7.** Let $d$ represent a person's distance, in feet, from the fulcrum of a seesaw, and $w$ represent the person's weight, in pounds.

To balance the seesaw, $d$ and $w$ are inversely proportional:

$$dw = k$$

Substitute $d = 4$ and $w = 150$ to find the constant of proportionality, $k$:

$$(4)(150) = k$$

$$600 = k$$

Using $k = 600$, substitute $w = 120$ and solve for the distance, $d$:

$$d(120) = 600$$

$$d = 5$$

Dan should sit 5 feet from the fulcrum to balance the seesaw.

The correct choice is **(4)**.

**8.** The *last* term in the expansion of $(x + 2y)^5$ is $(2y)^5 = 32y^5$.

The correct choice is **(4)**.

**9.** The given equation is:

$$\log_x 4 + \log_x 9 = 2$$

since $\log A + \log B = \log AB$:

$$\log_x(4 \cdot 9) = 2$$

$$\log_x 36 = 2$$

$$x^2 = 36$$

Rewrite the logarithm equation as an equivalent exponential equation, and solve for $x$:

$$x = 6$$

The correct choice is **(2)**.

**10.** The given sum is:

$$\frac{1}{\sqrt{3}} + \frac{1}{\sqrt{2}}$$

Rewrite each term as an equivalent fraction with a rational denominator:

$$\frac{1}{\sqrt{3}} \cdot \frac{\sqrt{3}}{\sqrt{3}} + \frac{1}{\sqrt{2}} \cdot \frac{\sqrt{2}}{\sqrt{2}}$$

Simplify:

$$\frac{\sqrt{3}}{3} + \frac{\sqrt{2}}{2}$$

Rewrite each term as an equivalent fraction with 6 as the common denominator:

$$\frac{\sqrt{3}}{3} \cdot \frac{2}{2} + \frac{\sqrt{2}}{2} \cdot \frac{3}{3}$$

$$\frac{2\sqrt{3}}{6} + \frac{3\sqrt{2}}{6}$$

Combine the fractions by writing the sum of their numerators over the common denominator:

$$\frac{2\sqrt{3} + 3\sqrt{2}}{6}$$

The correct choice is (**1**).

**11.** Choice (3), the quadratic equation $x^2 + x + 1 = 0$, is written in standard form as $ax^2 + bx + c = 0$ with $a = 1$, $b = 1$, and $c = 1$.

The nature of the roots can be determined by evaluating the discriminant:

$$b^2 - 4ac = (1)^2 - 4(1)(1) = -3$$

The discriminant is the expression under the square root in the quadratic formula. Because $-3$ is negative, the roots of this equation are imaginary.

The correct choice is (**3**).

**12.** The given equation is:

$$\log k = c \log v + \log p$$

The product of a constant and the logarithm of a number is the logarithm of the number raised to the constant power:

$$\log k = \log v^c + \log p$$

The sum of the logarithms of two numbers is the logarithm of the product:

$$\log k = \log v^c p$$

If the logarithms of two numbers are equal, the numbers must be equal:

$$k = v^c p$$

The correct choice is (**1**).

**13.** The given expression, $\displaystyle\sum_{r=1}^{3} {}_4C_r$, represents the sum of the terms ${}_4C_r$ for integer values of $r$ from 1 to 3. To evaluate the summation, substitute each of the values $r = 1$, 2, and 3 into the expression and add the results obtained from your calculator:

$$\sum_{r=1}^{3} {}_4C_r = {}_4C_1 + {}_4C_2 + {}_4C_3 = 4 + 6 + 4 = 14$$

The correct choice is (**2**).

**14.** It is given that, in $\triangle ABC$, $m\angle A = 30$, $m\angle C = 50$, and $AC = 13$, as shown in the accompanying diagram.

Since the sum of the degree measures of the angles of any triangle must be 180, $m\angle B = 180 - 30 - 50 = 100$.

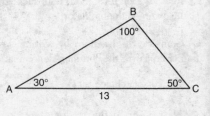

Find the length of side $\overline{AB}$ using the Law of Sines and your calculator:

$$\frac{\sin 50}{AB} = \frac{\sin 100}{13}$$
$$AB \sin 100 = 13 \sin 50$$
$$AB = \frac{13 \sin 50}{\sin 100}$$
$$\approx 10.112$$

To the *nearest tenth*, the length of side $\overline{AB}$ is 10.1.

The correct choice is (**2**).

**15.** The given expression is:

Since $i^4 = 1$, rewrite each term in terms of $i^4$ and simplify:

Substitute $i^2 = -1$, and combine like terms to simplify the expression:

$$i^{16} + i^6 - 2i^5 + i^{13}$$
$$\left(i^4\right)^4 + i^4 \cdot i^2 - 2i^4 \cdot i + \left(i^4\right)^3 i$$
$$1 + i^2 - 2i + i$$
$$1 + (-1) - 2i + i$$
$$-i$$

The correct choice is (**4**).

**16.** It is given that point $(a,b)$ lies on the graph $y = \mathrm{f}(x)$. To find the corresponding point on the graph $y = \mathrm{f}^{-1}(x)$, the inverse of this function, use the transformation $(x,y) \rightarrow (y,x)$ to reverse the ordered pair. Then the graph $y = \mathrm{f}^{-1}(x)$ must contain point $(b,a)$.

The correct choice is (**1**).

**17.** When a quadratic equation is written in the standard form $ax^2 + bx + c = 0$, the sum of the roots is $-\dfrac{b}{a}$ and the product of the roots is $\dfrac{c}{a}$.

For the given equation, $x^2 + 3x - 5 = 0$, $a = 1$, $b = 3$, and $c = -5$. Hence the sum of the roots is $-\dfrac{b}{a} = -\dfrac{3}{1} = -3$, and the product of the roots is $\dfrac{c}{a} = \dfrac{-5}{1} = -5$.

If the sum of the roots is added to the product of the roots, the result is $(-3) + (-5) = -8$.

The correct choice is **(4)**.

**18.** The expression $\dfrac{3^{\frac{1}{3}}}{3^{-\frac{2}{3}}} = 3^{\frac{1}{3}} \cdot 3^{\frac{2}{3}} = 3^{\frac{1}{3} + \frac{2}{3}} = 3^1 = 3$.

The correct choice is **(3)**.

**19.** The graph of figure $\text{I}$ is given in the accompanying diagram.

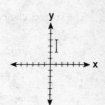

To find the image of $\text{I}$ after a transformation defined by $r_{y=x} \circ R_{90°}$, first rotate figure $\text{I}$ 90° counterclockwise, as shown in the second diagram.

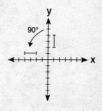

Now reflect the image of $\text{I}$ across the line $y = x$, as shown in the third diagram.

The final image, which represents $\text{I}$ after a transformation defined by $r_{y=x} \circ R_{90°}$, is shown in graph (3).

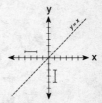

The correct choice is **(3)**.

**20.** The given expression is:

$$\frac{\dfrac{x}{x+2}}{1-\dfrac{x}{x+2}}$$

To simplify this complex fraction, multiply the numerator and denominator by the common denominator of the fraction terms, $x + 2$:

$$\frac{\dfrac{x}{x+2}\left(\dfrac{x+2}{1}\right)}{1\left(\dfrac{x+2}{1}\right)-\dfrac{x}{x+2}\left(\dfrac{x+2}{1}\right)}$$

In each term cancel the common factors, then simplify:

$$\frac{\dfrac{x}{\cancel{x+2}}\left(\dfrac{\cancel{x+2}}{1}\right)}{1\left(\dfrac{x+2}{1}\right)-\dfrac{x}{\cancel{x+2}}\left(\dfrac{\cancel{x+2}}{1}\right)}$$

$$\frac{x}{x+2-x}$$

$$\frac{x}{2}$$

The correct choice is **(2)**.

# PART II

**21.** It is given that a used car depreciates 13% of its value each year. Hence the value of the car each year is $100\% - 13\% = 87\%$ of its value the preceding year.

In 1999 the value of the car was \$11900.

In 2000 the value of the car was $11900(0.87)$.

In 2001 the value of the car was $11900(0.87)(0.87) = 11900(0.87)^2$.

In 2002 the value of the car is $11900(0.87)^2(0.87) = 11900(0.87)^3 \approx \$7836.19$.

The value of the car in July 2002, to the *nearest hundred dollars*, is **7,800**.

**22.** It is given that the amount of time a teenager plays video games in any given week is normally distributed with a mean of 15 hours per week and a standard deviation of 3 hours. To determine the probability that a teenager plays video games between 15 and 18 hours a week, first determine how far, in standard deviation units, each of these times is from the mean.

- 15 is the mean; hence it is 0 standard deviations from the mean.
- 18 is $\dfrac{18 - 15}{3} = 1$ standard deviation above the mean.

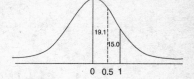

The accompanying figure shows that the percentage of scores between the mean and 1 standard deviation above the mean is approximately:

$19.1 + 15.0 = 34.1$.

The probability is approximately **0.341** or **34.1%**.

**23.** When six equally spaced points on the circumference of a circle are connected to its center, as shown in the accompanying diagram, the central angles formed are all congruent. Since the measure of the entire circle is $2\pi$  radians, the radian measure of each central angle is $\dfrac{1}{6}(2\pi)$.

In simplest form, the radian measure of the angle is $\dfrac{\pi}{3}$.

**24.** It is given that the Franklins will invest the amount $P = \$3,500$ at an interest rate of $r = 8.25\% = 0.0825$. Interest is compounded monthly, or $n = 12$ times per year, for $t = 5$ years.

Substitute these values in the given compound interest formula, and use your calculator to evaluate $A$:

$$A = P\left(1 + \frac{r}{n}\right)^{nt}$$

$$= 3500\left(1 + \frac{0.0825}{12}\right)^{12(5)}$$

$$\approx \$5,279.61$$

After 5 years the value of the account, in dollars, will be **5,279.61**.

**25.** It is given that a toy truck, $K$, is 4 feet from Alex and 3 feet from Dominic along chord $\overline{AD}$, as shown in the accompanying diagram. It is also given that the truck is 8 feet from Meira along chord $\overline{MT}$.

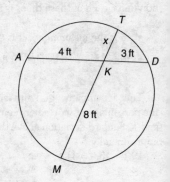

When two chords of a circle intersect, the product of the segments of one equals the product of the segments of the other:

$$(TK)(MK) = (AK)(DK)$$

Substitute the given distances, and solve:

$$(TK)(8) = (4)(3)$$

$$TK = 1.5$$

To the *nearest tenth of a foot*, Tamara is **1.5** from the truck.

**26.** It is given that two sides of a triangular-shaped pool measure 16 feet and 21 feet, and the included angle measures 58°, as shown in the accompanying diagram.

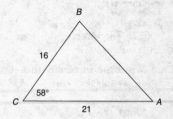

The area of a triangle is one-half the product of any two sides and the sine of the included angle:

Substitute the given measurements, and use your calculator to evaluate the area:

$$K = \frac{1}{2}\,ab\sin C$$

$$= \frac{1}{2}(16)(21)\sin 58°$$

$$\approx 142.472$$

To the *nearest tenth of a square foot*, the area is **142.5**.

# PART III

**27.** The given expressions for profit and cost are:

$$P = 500x, C = \frac{3200000}{x} + 60000$$

Write the inequality that represents profit greater than cost:

$$500x > \frac{3200000}{x} + 60000$$

Since the number of calculators is greater than 0, multiply each term in the inequality by $x$ and simplify:

$$500x(x) > \frac{3200000}{x}(x) + 60000(x)$$

$$500x^2 > 3200000 + 60000x$$

Rewrite the quadratic inequality in the standard form $ax^2 + bx + c > 0$:

$$500x^2 - 60000x - 3200000 > 0$$

Divide each term by 100:

$$5x^2 - 600x - 32000 > 0$$

The roots of the corresponding quadratic equality can be found by using the quadratic formula:

$$x = \frac{-b \pm \sqrt{b^2 - 4ac}}{2a}$$

Substitute $a = 5$, $b = -600$, and $c = -32,000$, and solve for $x$:

$$= \frac{-(-600) \pm \sqrt{(-600)^2 - 4(5)(-32000)}}{2(5)}$$

$$= \frac{600 \pm \sqrt{1000000}}{10}$$

$$x = \frac{600 + 1000}{10} \text{ or } x = \frac{600 - 1000}{10}$$

The number of calculators must be positive, so eliminate the negative root:

$$= 161 \qquad \text{or} \qquad = -40$$

Profit is greater than cost when a minimum of **161** calculators is sold.

**28.** It is given that forces of 1,500 pounds and 2,000 pounds are applied, and the resultant force is 3,000 pounds, as shown in the accompanying diagram. Since opposite sides of a parallelogram are congruent, $AR = CB = 1,500$.

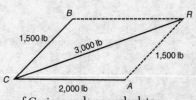

To find the measure of $\angle A$, use the Law of Cosines and your calculator:

$$a^2 = b^2 + c^2 - 2bc \cos A$$

$$3000^2 = 1500^2 + 2000^2 - 2(1500)(2000) \cos A$$

$$2(1500)(2000) \cos A = 1500^2 + 2000^2 - 3000^2$$

$$\cos A = \frac{1500^2 + 2000^2 - 3000^2}{2(1500)(2000)}$$

$$\cos A = -0.458\overline{3}$$

$$A = \cos^{-1}(-0.458\overline{3})$$

$$\approx 117.28°$$

In a parallelogram, consecutive angles are supplementary. Hence m∠BCA = 180° − 117.28° = 62.72°.

To the *nearest degree*, the angle between the two applied forces is **63**.

**29.** <u>Algebraic Solution:</u> Let $h$ represent the rock's height (in meters) above the ground $t$ seconds after it is thrown. It is given that $h = 2 + 24t - 4.9t^2$, which is a quadratic equation. The graph of a quadratic equation of the form $y = at^2 + bt + c$ is a parabola. The maximum height of the rock is at the vertex (turning point) of this parabola. The vertex lies on the axis of symmetry, whose equation is $t = -\dfrac{b}{2a}$. To find the value of $t$ at the turning point substitute $a = -4.9$ and $b = 24$:

$$t = -\frac{b}{2a} = -\frac{24}{2(-4.9)} \approx 2.449 \text{ seconds.}$$

To the *nearest hundredth of a second*, the number of seconds after the rock is thrown that it reaches maximum height is **2.45**.

To find the maximum height of the rock, substitute $t = 2.45$ in the equation for height:

$$h = 2 + 24(2.45) - 4.9(2.45)^2 \approx 31.388 \text{ meters.}$$

To the *nearest hundredth of a meter*, the maximum height that the rock will reach is **31.39**.

The rock hits the ground when $h = 0$:

$$2 + 24t - 4.9t^2 = 0$$

Rewrite this quadratic equation in the standard form $at^2 + bt + c = 0$:

$$-4.9t^2 + 24t + 2 = 0$$

To find the time when the rock hits the ground, use the quadratic formula and your calculator:

$$t = \frac{-b \pm \sqrt{b^2 - 4ac}}{2a}$$

Substitute $a = -4.9$, $b = 24$, and $c = 2$, and solve for $t$:

$$= \frac{-24 \pm \sqrt{24^2 - 4(-4.9)(2)}}{2(-4.9)}$$

$$= \frac{-24 \pm \sqrt{615.2}}{-9.8}$$

The time must be positive, so eliminate the negative root:

$$t \approx \frac{-24 + 24.803}{-9.8} \quad \text{or} \quad t = \frac{-24 - 24.803}{-9.8}$$

$$\approx -0.082 \quad\quad \text{or} \quad \approx 4.9799$$

To the *nearest hundredth of a second*, the number of seconds after the rock is thrown that it hits the ground is **4.98**.

<u>Graphical Solution</u>: Let $y$ represent the rock's height; then the given function is $y = 2 + 24t - 4.9t^2$. Using your graphing calculator, graph this function in the graphing window with $-2 \leq t \leq 6$ and $-10 \leq y \leq 40$, as shown in the accompanying diagram.

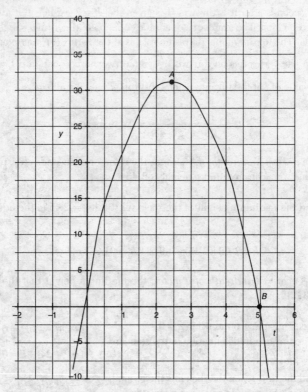

To determine when and where the rock will reach its maximum height, find the coordinates of $A$, the turning point of the graph. Use your graphing calculator to determine that the coordinates of $A$ are approximately (2.449,31.388).

To the *nearest hundredth*, **2.45** seconds after the rock is thrown it will reach its maximum height of **31.39** meters.

To determine when the rock will hit the ground, find the coordinates of $B$, the point where the graph of the curve intersects the $x$-axis. Use your graphing calculator to determine that the coordinates of $B$ are approximately (4.9799,0).

To the *nearest hundredth of a second*, the number of seconds after the rock is thrown that it hits the ground is **4.98**.

**30.** It is given that △ABC is *not* isosceles. Use an indirect proof to show that altitude $\overline{BD}$, if drawn as shown in the accompanying diagram, does *not* bisect $\overline{AC}$.

Given: △ABC is *not* isosceles.
$\overline{BD}$ is an altitude.

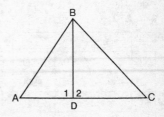

Prove: $\overline{BD}$ does *not* bisect $\overline{AC}$.

## PROOF

| Statement | Reason |
|---|---|
| 1. △ABC is *not* isosceles. $\overline{BD}$ is an altitude. | 1. Given |
| 2. Suppose $\overline{BD}$ bisects $\overline{AC}$. | 2. Assumption for indirect proof |
| 3. $\overline{AD} \cong \overline{CD}$ (Side) | 3. A bisector divides a segment into two congruent segments. |
| 4. $\overline{BD} \perp \overline{AC}$ | 4. An altitude of a triangle is drawn from a vertex perpendicular to the opposite side. |
| 5. ∠1 and ∠2 are right angles. | 5. Perpendicular lines form right angles. |
| 6. ∠1 ≅ ∠2 (Angle) | 6. All right angles are congruent. |
| 7. $\overline{BD} \cong \overline{BD}$ (Side) | 7. Congruence is reflexive. |
| 8. △ABD ≅ △CBD | 8. SAS ≅ SAS |
| 9. $\overline{AB} \cong \overline{CB}$ | 9. CPCTC |
| 10. △ABC is isosceles. (*CONTRADICTION*) | 10. A triangle with two congruent sides is isosceles. |
| 11. $\overline{BD}$ does *not* bisect $\overline{AC}$. | 11. The contradiction shows that the assumption was false. |

**31.** *a.* To graph the equation $y = x^2$, first make a table of values for $-3 \leq x \leq 3$:

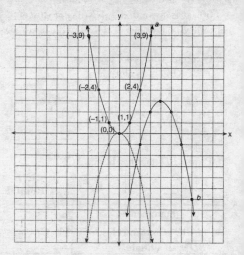

| $x$ | $y = x^2$ | $(x,y)$ |
|-----|-----------|---------|
| $-3$ | $y = (-3)^2 = 9$ | $(-3,9)$ |
| $-2$ | $y = (-2)^2 = 4$ | $(-2,4)$ |
| $-1$ | $y = (-1)^2 = 1$ | $(-1,1)$ |
| $0$ | $y = (0)^2 = 0$ | $(0,0)$ |
| $1$ | $y = (1)^2 = 1$ | $(1,1)$ |
| $2$ | $y = (2)^2 = 4$ | $(2,4)$ |
| $3$ | $y = (3)^2 = 9$ | $(3,9)$ |

Plot each of the points and connect them with a smooth curve. Label the parabola *a*, as shown in the accompanying diagram.

*b.* Graph the equation $y = -(x - 4)^2 + 3$ by performing the following transformations on the graph of *a*: $y = x^2$:

- $r_{x\text{-axis}}$: Reflect the graph of *a* across the *x*-axis.
- $T_{(4,3)}$: Shift the reflection right 4 units and up 3 units.

Graph *b* is graph *a* after the transformation $T_{(4,3)} \circ r_{x\text{-axis}}$, as shown in the diagram.

**32.** To construct a scatterplot of the given data, plot point $(d,y)$ for each of cable diameters given, as shown in the accompanying diagram.

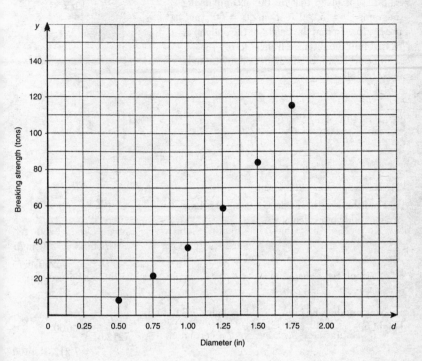

Enter the given data into your calculator with diameter as $x$ and breaking strength as $y$, and use the calculator to find the exponential regression equation.

The equation, with the regression coefficients rounded to the *nearest tenth*, is $y = 4.8(6.8)^x$.

# PART IV

**33.** It is given that, in the accompanying diagram, $CJ = 5,280$ feet, m$\angle C = 60°$, and m$\angle J = 75°$. To find the height of the balloon at $B$, first find the distance from $C$ to $B$.

Since the sum of the angles of any triangle must be $180°$, m$\angle B = 180 - 60 - 75 = 45$.

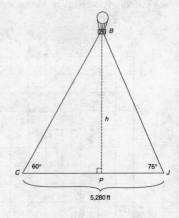

Find the length of side $\overline{CB}$ using the Law of Sines and your calculator:

$$\frac{\sin 75}{CB} = \frac{\sin 45}{5280}$$
$$CB \sin 45 = 5280 \sin 75$$
$$CB = \frac{5280 \sin 75}{\sin 45}$$
$$\approx 7,212.61$$

Now use the sine ratio in right triangle $CPB$ to find height $h$:

$$\frac{h}{7212.61} = \sin 60$$
$$h = 7,212.61 \sin 60$$
$$\approx 6,246.3$$

To the *nearest foot*, the height of the balloon is **6,246**.

**34.** The accompanying diagram shows electrical circuits hooked up in parallel.

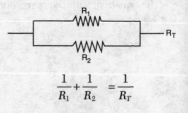

The given equation is:

$$\frac{1}{R_1} + \frac{1}{R_2} = \frac{1}{R_T}$$

It is also given that $R_1 = x$, $R_2 = x + 3$, and $R_T = 2.25$:

$$\frac{1}{x} + \frac{1}{x+3} = \frac{1}{2.25}$$

To solve an equation with fractions, first multiply both sides by the common denominator, here $2.25x(x + 3)$:

$$\left(\frac{1}{x}\right)2.25x(x+3)+\left(\frac{1}{x+3}\right)2.25x(x+3) = \left(\frac{1}{2.25}\right)2.25x(x+3)$$

In each term, cancel the common factors:

$$\left(\frac{1}{\cancel{x}}\right)2.25\cancel{x}(x+3)+\left(\frac{1}{\cancel{x+3}}\right)2.25x(\cancel{x+3}) = \left(\frac{1}{\cancel{2.25}}\right)\cancel{2.25}x(x+3)$$

Simplify:

$$2.25(x+3)+2.25(x) = x(x+3)$$
$$2.25x+6.75+2.25x = x^2+3x$$
$$4.5x+6.75 = x^2+3x$$

Rewrite the quadratic equation in the standard form $ax^2 + bx + c = 0$:

$$x^2-1.5x-6.75=0$$

Solve this equation using the quadratic formula:

$$x = \frac{-b\pm\sqrt{b^2-4ac}}{2a}$$

Substitute $a = 1$, $b = -1.5$, and $c = -6.75$, and simplify:

$$= \frac{-(-1.5)\pm\sqrt{(-1.5)^2-4(1)(-6.75)}}{2(1)}$$

$$= \frac{1.5\pm\sqrt{29.25}}{2}$$

Use your calculator to find the positive value of $x$:

$$= \frac{1.5+\sqrt{29.25}}{2}\approx 3.454$$

To the *nearest tenth of an ohm*, the positive value of $R_1$ is **3.5**.

| Topic | Question Numbers | Number of Points | Your Points |
|---|---|---|---|
| 1. Properties of Real Numbers | — | — | |
| 2. Sequences | 21 | 2 | |
| 3. Complex Numbers | 15 | 2 | |
| 4. Inequalities, Absolute Value | 3 | 2 | |
| 5. Algebraic Expressions, Fractions | 8, 20, 34 | 2 + 2 + 3 = 7 | |
| 6. Exponents (zero, negative, rational, scientific notation) | 18 | 2 | |
| 7. Radical Expressions | 10 | 2 | |
| 8. Quadratic Equations (factors, formula, discriminant) | 11, 17, 27, 29, 34 | 2 + 2 + 4 + 2 + 3 = 13 | |
| 9. Systems of Equations | — | — | |
| 10. Functions (graphs, domain, range, roots) | 4 | 2 | |
| 11. Inverse Functions, Composition | 5, 16 | 2 + 2 = 4 | |
| 12. Linear Functions | — | — | |
| 13. Parabolas (max/min, axis of symmetry) | 29 | 2 | |
| 14. Hyperbolas (including inverse variation) | 7 | 2 | |
| 15. Ellipses | 6 | 2 | |
| 16. Exponents and Logarithms | 9, 12, 24 | 2 + 2 + 2 = 6 | |
| 17. Trig. (circular functions, unit circle, radians) | 23 | 2 | |
| 18. Trig. Equations and Identities | — | — | |
| 19. Solving Triangles (sin/cos laws, Pythag. thm., ambig. case) | 14, 26, 28, 33 | 2 + 2 + 4 + 6 = 14 | |
| 20. Coordinate Geom. (slope, distance, midpoint, circle) | — | — | |
| 21. Transformations, Symmetry | 19, 31 | 2 + 4 = 6 | |
| 22. Circle Geometry | 25 | 2 | |
| 23. Congruence, Similarity, Proportions | — | — | |
| 24. Probability (including Bernoulli events) | 1 | 2 | |
| 25. Normal Curve | 2, 22 | 2 + 2 = 4 | |
| 26. Statistics (center, spread, summation notation) | 13 | 2 | |
| 27. Correlation, Modeling, Prediction | 32 | 4 | |
| 28. Algebraic Proofs, Word Problems | — | — | |
| 29. Geometric Proofs | 30 | 4 | |
| 30. Coordinate Geometry Proofs | — | — | |

**Total Raw Score =** _____

## HOW TO CONVERT YOUR RAW SCORE
## TO YOUR MATH B REGENTS EXAMINATION SCORE

Below is the conversion chart that must be used to determine your final score on the August 2002 Regents Examination in Math B. To find your final exam score, locate in the column labeled "Raw Score" the total number of points you scored. Then locate in the adjacent column to the right the scaled score that corresponds to your raw score. The scaled score is your final Math B Regents Examination score.

### Regents Examination in Math B—August 2002
### Chart for Converting Total Test Raw Scores to
### Final Examination Scores (Scaled Scores)

| Raw Score | Scaled Score | Raw Score | Scaled Score | Raw Score | Scaled Score |
|---|---|---|---|---|---|
| 88 | 100 | 58 | 80 | 28 | 54 |
| 87 | 99 | 57 | 79 | 27 | 53 |
| 86 | 98 | 56 | 78 | 26 | 52 |
| 85 | 97 | 55 | 78 | 25 | 50 |
| 84 | 96 | 54 | 77 | 24 | 49 |
| 83 | 96 | 53 | 77 | 23 | 48 |
| 82 | 95 | 52 | 76 | 22 | 46 |
| 81 | 94 | 51 | 75 | 21 | 45 |
| 80 | 93 | 50 | 75 | 20 | 43 |
| 79 | 93 | 49 | 74 | 19 | 41 |
| 78 | 92 | 48 | 73 | 18 | 40 |
| 77 | 91 | 47 | 72 | 17 | 38 |
| 76 | 90 | 46 | 72 | 16 | 36 |
| 75 | 90 | 45 | 71 | 15 | 34 |
| 74 | 89 | 44 | 70 | 14 | 33 |
| 73 | 88 | 43 | 69 | 13 | 31 |
| 72 | 88 | 42 | 69 | 12 | 29 |
| 71 | 87 | 41 | 68 | 11 | 27 |
| 70 | 87 | 40 | 67 | 10 | 24 |
| 69 | 86 | 39 | 66 | 9 | 22 |
| 68 | 85 | 38 | 65 | 8 | 20 |
| 67 | 85 | 37 | 64 | 7 | 18 |
| 66 | 84 | 36 | 63 | 6 | 15 |
| 65 | 84 | 35 | 62 | 5 | 13 |
| 64 | 83 | 34 | 61 | 4 | 11 |
| 63 | 82 | 33 | 60 | 3 | 8 |
| 62 | 82 | 32 | 59 | 2 | 5 |
| 61 | 81 | 31 | 58 | 1 | 3 |
| 60 | 81 | 30 | 57 | 0 | 0 |
| 59 | 80 | 29 | 56 | | |

# Examination June 2003

## Math B

## FORMULAS

### Area of Triangle

$$K = \frac{1}{2}ab\sin C$$

### Function of the Sum of Two Angles

$$\sin(A + B) = \sin A \cos B + \cos A \sin B$$
$$\cos(A + B) = \cos A \cos B - \sin A \sin B$$

### Function of the Difference of Two Angles

$$\sin(A - B) = \sin A \cos B - \cos A \sin B$$
$$\cos(A - B) = \cos A \cos B + \sin A \sin B$$

### Law of Sines

$$\frac{a}{\sin A} = \frac{b}{\sin B} = \frac{c}{\sin C}$$

### Law of Cosines

$$a^2 = b^2 + c^2 - 2bc \cos A$$

### Functions of the Double Angle

$$\sin 2A = 2 \sin A \cos A$$
$$\cos 2A = \cos^2 A - \sin^2 A$$
$$\cos 2A = 2 \cos^2 A - 1$$
$$\cos 2A = 1 - 2 \sin^2 A$$

## Functions of the Half Angle

$$\sin\frac{1}{2}A = \pm\sqrt{\frac{1-\cos A}{2}}$$

$$\cos\frac{1}{2}A = \pm\sqrt{\frac{1+\cos A}{2}}$$

## Normal Curve
## Standard Deviation

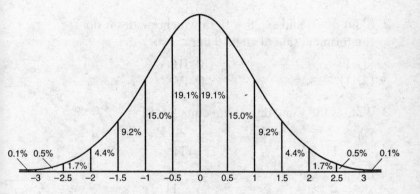

# PART I

Answer all questions in this part. Each correct answer will receive 2 credits. No partial credit will be allowed. For each question, write in the spaces provided the numeral preceding the word or expression that best completes the statement or answers the question.   [40]

1  For which value of $x$ is $y = \log x$ undefined?

    (1) 0                    (3) $\pi$

    (2) $\dfrac{1}{10}$            (4) 1.483           1 _____

2  If $\sin \theta > 0$ and $\sec \theta < 0$, in which quadrant does the terminal side of angle $\theta$ lie?

    (1) I                  (3) III

    (2) II                (4) IV           2 _____

3  What is the value of $x$ in the equation $81^{x+2} = 27^{5x+4}$?

    (1) $-\dfrac{2}{11}$          (3) $\dfrac{4}{11}$

    (2) $-\dfrac{3}{2}$           (4) $-\dfrac{4}{11}$       3 _____

4  The relationship between voltage, $E$, current, $I$, and resistance, $Z$, is given by the equation $E = IZ$. If a circuit has a current $I = 3 + 2i$ and a resistance $Z = 2 - i$, what is the voltage of this circuit?

    (1) $8 + i$            (3) $4 + i$

    (2) $8 + 7i$          (4) $4 - i$         4 _____

5  Which expression is equivalent to $\dfrac{4}{3+\sqrt{2}}$?

    (1) $\dfrac{12+4\sqrt{2}}{7}$       (3) $\dfrac{12-4\sqrt{2}}{7}$

    (2) $\dfrac{12+4\sqrt{2}}{11}$     (4) $\dfrac{12-4\sqrt{2}}{11}$     5 _____

6 What are the coordinates of point *P*, the image of point (3,–4) after a reflection in the line *y* = *x*?

(1) (3,4)　　　　　　(3) (4,–3)

(2) (–3,4)　　　　　(4) (–4,3)　　　　6 ____

7 The roots of the equation $ax^2 + 4x = -2$ are real, rational, and equal when *a* has a value of

(1) 1　　　　　　　(3) 3

(2) 2　　　　　　　(4) 4　　　　　7 ____

8 Two objects are $2.4 \times 10^{20}$ centimeters apart. A message from one object travels to the other at a rate of $1.2 \times 10^5$ centimeters per second. How many seconds does it take the message to travel from one object to the other?

(1) $1.2 \times 10^{15}$　　　　(3) $2.0 \times 10^{15}$

(2) $2.0 \times 10^4$　　　　　(4) $2.88 \times 10^{25}$　　　8 ____

9 If f(*x*) = cos *x*, which graph represents f(*x*) under the composition $r_{y\text{-axis}} \circ r_{x\text{-axis}}$?

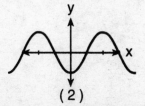

(1)

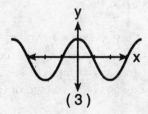

(3)

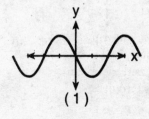

(2)

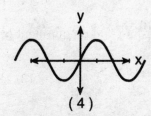

(4)　　　　9 ____

10 Which diagram represents a relation in which each member of the domain corresponds to only one member of its range?

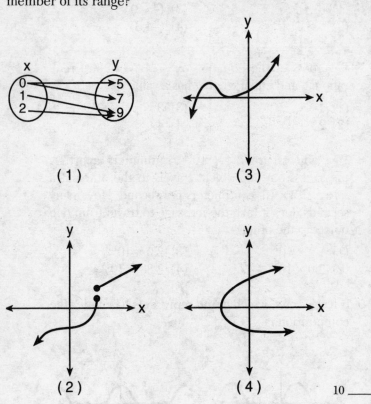

( 1 )                              ( 3 )

( 2 )                              ( 4 )

10 ____

11 The accompanying diagram represents the elliptical path of a ride at an amusement park.

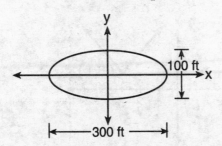

Which equation represents this path?

(1) $x^2 + y^2 = 300$

(3) $\dfrac{x^2}{150^2} + \dfrac{y^2}{50^2} = 1$

(2) $y = x^2 + 100x + 300$

(4) $\dfrac{x^2}{150^2} - \dfrac{y^2}{50^2} = 1$    11 _____

12 If $A$ and $B$ are positive acute angles, $\sin A = \dfrac{5}{13}$, and $\cos B = \dfrac{4}{5}$, what is the value of $\sin (A + B)$?

(1) $\dfrac{56}{65}$

(3) $\dfrac{33}{65}$

(2) $\dfrac{63}{65}$

(4) $-\dfrac{16}{65}$    12 _____

13 Which transformation is an opposite isometry?

(1) dilation            (3) rotation of 90°

(2) line reflection     (4) translation    13 _____

14 Which equation is represented by the accompanying graph?

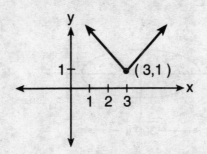

(1) $y = |x| - 3$          (3) $y = |x + 3| - 1$
(2) $y = (x - 3)^2 + 1$      (4) $y = |x - 3| + 1$     14 _____

15 What is the value of $i^{99} - i^3$?

(1) 1                (3) $-i$
(2) $i^{96}$            (4) 0                  15 _____

16 If $\log a = 2$ and $\log b = 3$, what is the numerical value of $\log \dfrac{\sqrt{a}}{b^3}$?

(1) 8             (3) 25
(2) −8          (4) −25             16 _____

17 In simplest form, $\dfrac{\dfrac{1}{x^2} - \dfrac{1}{y^2}}{\dfrac{1}{y} + \dfrac{1}{x}}$ is equal to

(1) $\dfrac{x - y}{xy}$          (3) $x - y$

(2) $\dfrac{y - x}{xy}$          (4) $y - x$        17 _____

18 What is the solution set of the inequality $|3 - 2x| \geq 4$?

(1) $\left\{ x | \dfrac{7}{2} \leq x \leq -\dfrac{1}{2} \right\}$      (3) $\left\{ x | x \leq -\dfrac{1}{2} \text{ or } x \geq \dfrac{7}{2} \right\}$

(2) $\left\{ x | -\dfrac{1}{2} \leq x \leq \dfrac{7}{2} \right\}$      (4) $\left\{ x | x \leq \dfrac{7}{2} \text{ or } x \geq -\dfrac{1}{2} \right\}$    18 \_\_\_\_

19 What value of $x$ in the interval $0° \leq x \leq 180°$ satisfies the equation $\sqrt{3} \tan x + 1 = 0$?

(1) −30°      (3) 60°
(2) 30°      (4) 150°    19 \_\_\_\_

20 In the accompanying diagram, $\overline{CA} \perp \overline{AB}$, $\overline{ED} \perp \overline{DF}$, $\overline{ED} \parallel \overline{AB}$, $\overline{CE} \cong \overline{BF}$, $\overline{AB} \cong \overline{ED}$, and m$\angle CAB$ = m$\angle FDE$ = 90.

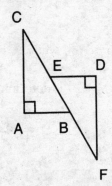

Which statement would *not* be used to prove $\triangle ABC \cong \triangle DEF$?

(1) SSS $\cong$ SSS      (3) AAS $\cong$ AAS
(2) SAS $\cong$ SAS      (4) HL $\cong$ HL    20 \_\_\_\_

## PART II

Answer all questions in this part. Each correct answer will receive 2 credits. Clearly indicate the necessary steps, including appropriate formula substitutions, diagrams, graphs, charts, etc. For all questions in this part, a correct numerical answer with no work shown will receive only 1 credit.   [12]

21   Vanessa throws a tennis ball in the air. The function $h(t) = -16t^2 + 45t + 7$ represents the distance, in feet, that the ball is from the ground at any time $t$. At what time, to the *nearest tenth of a second*, is the ball at its maximum height?

22   If $f(x) = 2^x - 1$ and $g(x) = x^2 - 1$, determine the value of $(f \circ g)(3)$.

23 When air is pumped into an automobile tire, the pressure is inversely proportional to the volume. If the pressure is 35 pounds when the volume is 120 cubic inches, what is the pressure, in pounds, when the volume is 140 cubic inches?

24 In a certain school district, the ages of all new teachers hired during the last 5 years are normally distributed. Within this curve, 95.4% of the ages, centered about the mean, are between 24.6 and 37.4 years. Find the mean age and the standard deviation of the data.

25 Express the following rational expression in simplest form:

$$\frac{9-x^2}{10x^2-28x-6}$$

26 Evaluate: $2\sum_{n=1}^{5}(2n-1)$

## PART III

**Answer all questions in this part. Each correct answer will receive 4 credits. Clearly indicate the necessary steps, including appropriate formula substitutions, diagrams, graphs, charts, etc. For all questions in this part, a correct numerical answer with no work shown will receive only 1 credit.** [24]

27 The coordinates of quadrilateral $ABCD$ are $A(-1,-5)$, $B(8,2)$, $C(11,13)$, and $D(2,6)$. Using coordinate geometry, prove that quadrilateral $ABCD$ is a rhombus. [The use of the accompanying grid is optional.]

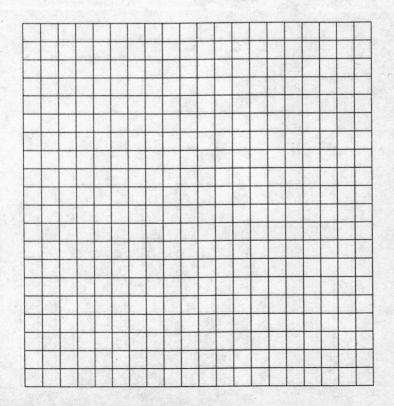

28 The price of a stock, A($x$), over a 12-month period decreased and then increased according to the equation A($x$) = $0.75x^2 - 6x + 20$, where $x$ equals the number of months. The price of another stock, B($x$), increased according to the equation B($x$) = $2.75x + 1.50$ over the same 12-month period. Graph and label both equations on the accompanying grid. State all prices, to the *nearest dollar*, when both stock values were the same.

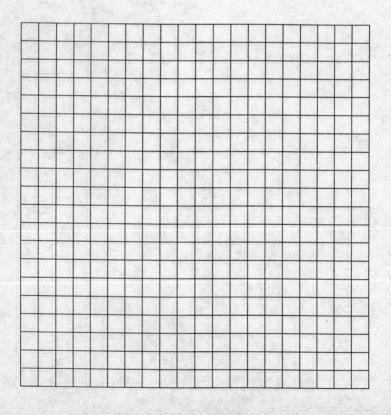

29 A pair of figure skaters graphed part of their routine on a grid. The male skater's path is represented by the equation $m(x) = 3 \sin\frac{1}{2}x$, and the female skater's path is represented by the equation $f(x) = -2 \cos x$. On the accompanying grid, sketch both paths and state how many times the paths of the skaters intersect between $x = 0$ and $x = 4\pi$.

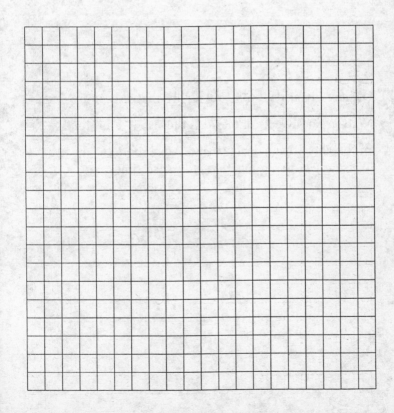

30 Sean invests \$10,000 at an annual rate of 5% compounded continuously, according to the formula $A = Pe^{rt}$, where $A$ is the amount, $P$ is the principal, $e = 2.718$, $r$ is the rate of interest, and $t$ is time, in years.

Determine, to the *nearest dollar*, the amount of money he will have after 2 years.

Determine how many years, to the *nearest year*, it will take for his initial investment to double.

31 On any given day, the probability that the entire Watson family eats dinner together is $\frac{2}{5}$. Find the probability that, during any 7-day period, the Watsons eat dinner together *at least* six times.

32 While sailing a boat offshore, Donna sees a lighthouse and calculates that the angle of elevation to
the top of the lighthouse is 3°, as shown in the
accompanying diagram. When she sails her boat
700 feet closer to the lighthouse, she finds that the
angle of elevation is now 5°. How tall, to the *nearest
tenth of a foot*, is the lighthouse?

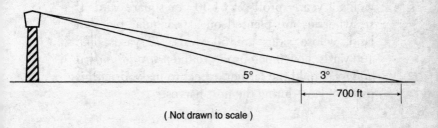

5°    3°

|← 700 ft →|

( Not drawn to scale )

## PART IV

**Answer all questions in this part. Each correct answer will receive 6 credits. Clearly indicate the necessary steps, including appropriate formula substitutions, diagrams, graphs, charts, etc. For all questions in this part, a correct numerical answer with no work shown will receive only 1 credit.** [12]

33 A farmer has determined that a crop of strawberries yields a yearly profit of $1.50 per square yard. If strawberries are planted on a triangular piece of land, whose sides are 50 yards, 75 yards, and 100 yards, how much profit, to the *nearest hundred dollars,* would the farmer expect to make from this piece of land during the next harvest?

34 For a carnival game, John is painting two circles, V and M, on a square dartboard.

a On the accompanying grid, draw and label circle V, represented by the equation $x^2 + y^2 = 25$, and circle M, represented by the equation $(x - 8)^2 + (y + 6)^2 = 4$.

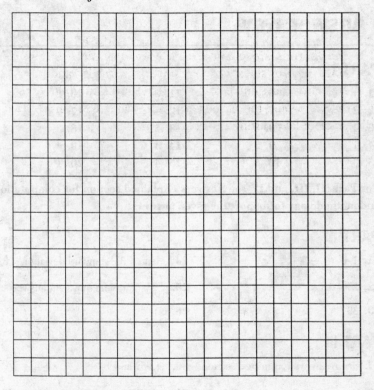

b A point, $(x,y)$, is randomly selected such that $-10 \leq x \leq 10$ and $-10 \leq y \leq 10$. What is the probability that point $(x,y)$ lies outside both circle V and circle M?

# Answers
# June 2003
## Math B

## Answer Key

### PART I

| | | | |
|---|---|---|---|
| **1.** 1 | **6.** 4 | **11.** 3 | **16.** 2 |
| **2.** 2 | **7.** 2 | **12.** 1 | **17.** 2 |
| **3.** 4 | **8.** 3 | **13.** 2 | **18.** 3 |
| **4.** 1 | **9.** 2 | **14.** 4 | **19.** 4 |
| **5.** 3 | **10.** 3 | **15.** 4 | **20.** 1 |

For Parts II, III, and IV, see **Answers Explained** section for computations and methodology to support the given answers.

### PART II

**21.** 1.4

**22.** 255

**23.** 30

**24.** Mean = 31; standard deviation = 3.2

**25.** $\dfrac{-x-3}{10x+2}$ or an equivalent answer

**26.** 50

### PART III

**27.** See **Answers Explained** section.

**28.** 9 and 26

**29.** Two

**30.** 11,052 and 14

**31.** $\dfrac{1,472}{78,125}$

**32.** 91.5

### PART IV

**33.** 2,700

**34.** *a* See **Answers Explained** section.

   *b* 0.7722345326 or an equivalent decimal answer

# Answers Explained

## PART I

**1.** The function $y = \log x$ is defined for $x > 0$. Therefore, it is undefined for $x = 0$.

The correct choice is **(1)**.

**2.** It is given that $\sin \theta > 0$ and $\sec \theta < 0$. The sine is positive for an angle that terminates in Quadrant I or II. Like the cosine, the secant is negative for an angle that terminates in Quadrant II or III. Hence, the terminal side of angle $\theta$ must lie in Quadrant II.

The correct choice is **(2)**.

**3.** The equation given is:

$$81^{x+2} = 27^{5x+4}$$

Rewrite the equation using a common base:

$$3^{4(x+2)} = 3^{3(5x+4)}$$

Set the exponents equal:

$$4(x+2) = 3(5x+4)$$

Solve the resulting equation for $x$:

$$4x + 8 = 15x + 12$$
$$-11x = 4$$
$$x = -\frac{4}{11}$$

The correct choice is **(4)**.

**4.** The equation given is:

$$E = IZ$$

To find the voltage, $E$, substitute the given current, $I = 3 + 2i$, and resistance, $Z = 2 - i$:

Multiply and simplify:

$$E = (3 + 2i)(2 - i)$$
$$= 6 + i - 2i^2$$
$$= 6 + i - 2(-1)$$
$$= 6 + i + 2$$
$$= 8 + i$$

The correct choice is **(1)**.

**5.** The given expression is:

$$\frac{4}{3 + \sqrt{2}}$$

Multiply the numerator and denominator by the conjugate of the denominator, $3 - \sqrt{2}$:

$$\frac{4}{3 + \sqrt{2}} \cdot \frac{3 - \sqrt{2}}{3 - \sqrt{2}}$$

$$\frac{4\left(3 - \sqrt{2}\right)}{\left(3 + \sqrt{2}\right)\left(3 - \sqrt{2}\right)}$$

$$\frac{12 - 4\sqrt{2}}{9 - 2}$$

$$\frac{12 - 4\sqrt{2}}{7}$$

The correct choice is **(3)**.

**6.** After a reflection in the line $y = x$, the coordinates of the image of point $(x,y)$ are $(y,x)$. Hence the coordinates of the image of point $(3,-4)$ are $(-4,3)$.

The correct choice is **(4)**.

**7.** The equation given is:

$$ax^2 + 4x = -2$$

Rewrite the equation in the standard form, $ax^2 + bx + c = 0$:

$$ax^2 + 4x + 2 = 0$$

It is given that the roots of the equation are real, rational, and equal. Hence, the value of the discriminant, $b^2 - 4ac$, must be 0:

$$b^2 - 4ac = 0$$

Substitute $b = 4$ and $c = 2$.

$$4^2 - 4a(2) = 0$$

Solve for $a$:

$$16 - 8a = 0$$
$$16 = 8a$$
$$2 = a$$

The correct choice is **(2)**.

**8.** To determine how many seconds the message takes to travel from one object to the other, divide the distance between the two objects by the rate at which the message travels:

$$\frac{2.4 \times 10^{20}\,\text{cm}}{1.2 \times 10^{5}\,\text{cm/s}} = 2.0 \times 10^{15}\,\text{s}$$

The correct choice is **(3)**.

**9.** The graph of $f(x) = \cos x$ is:

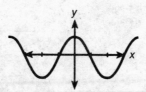

To graph $f$ under the composition $r_{y\text{-axis}} \circ r_{x\text{-axis}}$, first reflect the graph of $f$ across the $x$-axis, as shown in the accompanying diagram.

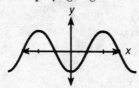

Now reflect that image across the $y$-axis. (Note that, because the first image was symmetric across the $y$-axis, this graph does not change.)

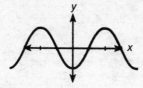

The correct choice is **(2)**.

**10.** Consider each of the choices offered:

- In choice (1) the domain value $x = 0$ corresponds to both $y = 5$ and $y = 7$; hence (1) is *not* correct.
- In choice (2) a vertical line can intersect the graph twice, indicating that a member of its domain (value of $x$) corresponds to two members of its range (values of $y$); hence (2) is *not* correct.
- In choice (3) no vertical line can intersect the graph more than once, indicating that each member of its domain corresponds to only one member of its range; hence (3) *is* correct.
- In choice (4) a vertical line can intersect the graph twice; hence (4) is *not* correct.

The correct choice is **(3)**.

**11.** An amusement park ride is in the shape of an ellipse with major axis 300 feet and minor axis 100 feet, as shown in the given diagram.

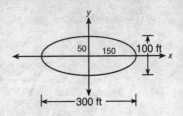

The equation of an ellipse with semimajor axis $a$ and semiminor axis $b$ is:

The semimajor axis of the given ellipse is 150, and the semiminor axis is 50. To find the equation that represents this path, substitute $a = 150$ and $b = 50$:

$$\frac{x^2}{a^2} + \frac{y^2}{b^2} = 1$$

$$\frac{x^2}{150^2} + \frac{y^2}{50^2} = 1$$

The correct choice is **(3)**.

**12.** It is given that $A$ is a positive acute angle, and that $\sin A = \frac{5}{13}$, as shown in the accompanying diagram. In this 5-12-13 right triangle, $\cos A = \frac{12}{13}$.

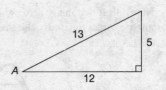

It is also given that $B$ is a positive acute angle, and that $\cos B = \frac{4}{5}$, as shown in the accompanying diagram. In this 3-4-5 right triangle, $\sin B = \frac{3}{5}$.

Then:

$$\sin(A + B) = \sin A \cos B + \cos A \sin B$$

$$= \frac{5}{13} \cdot \frac{4}{5} + \frac{12}{13} \cdot \frac{3}{5}$$

$$= \frac{56}{65}$$

The correct choice is **(1)**.

**13.** Dilations, rotations, and translations all preserve the orientation of a figure, but reflections do not. Therefore, line reflection is an opposite isometry.

The correct choice is (**2**).

**14.** The graph of the given function can be obtained by translating the graph of $y = |x|$ 3 units to the right and 1 unit upward, as shown in the accompanying graph. Hence the equation represented by the given graph is $y = |x - 3| + 1$.

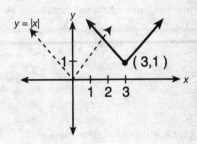

The correct choice is (**4**).

**15.** The given expression is:

$$i^{99} - i^3$$

Since $i^4 = 1$, rewrite each term in terms of $i^4$ and simplify:

$$(i^4)^{24} \cdot i^3 - i^3$$
$$(1)^{24} \cdot i^3 - i^3$$
$$i^3 - i^3$$
$$0$$

The correct choice is (**4**).

**16.** The given expression is:

$$\log \frac{\sqrt{a}}{b^3}$$

Since $\log\left(\dfrac{A}{B}\right) = \log A - \log B$:

$$\log a^{\frac{1}{2}} - \log b^3$$

The logarithm of a number raised to a constant power is the product of the constant and the logarithm of the number:

$$\tfrac{1}{2}\log a - 3\log b$$

It is given that $\log a = 2$ and $\log b = 3$:

$$\tfrac{1}{2}(2) - 3(3)$$
$$1 - 9$$
$$-8$$

The correct choice is (**2**).

**17.** The given expression is:

$$\dfrac{\dfrac{1}{x^2} - \dfrac{1}{y^2}}{\dfrac{1}{y} + \dfrac{1}{x}}$$

To simplify this complex fraction, multiply the numerator and denominator by the common denominator of the fraction terms, $x^2y^2$:

$$\dfrac{x^2y^2\left(\dfrac{1}{x^2} - \dfrac{1}{y^2}\right)}{x^2y^2\left(\dfrac{1}{y} + \dfrac{1}{x}\right)}$$

Distribute, and simplify:

$$\dfrac{x^2y^2\left(\dfrac{1}{x^2}\right) - x^2y^2\left(\dfrac{1}{y^2}\right)}{x^2y^2\left(\dfrac{1}{y}\right) + x^2y^2\left(\dfrac{1}{x}\right)}$$

$$\dfrac{y^2 - x^2}{x^2y + xy^2}$$

Factor the numerator and denominator:

$$\dfrac{(y - x)(y + x)}{xy(x + y)}$$

Since $(x + y) = (y + x)$, the fraction may be reduced by canceling the common factor:

$$\dfrac{y - x}{xy}$$

The correct choice is (**2**).

**18.** The given inequality is:

$$|3 - 2x| \geq 4$$

Rewrite the inequality without the absolute value expression:
Solve each inequality:

$$3 - 2x \leq -4 \quad \text{or} \quad 3 - 2x \geq 4$$
$$-2x \leq -7 \quad \text{or} \quad -2x \geq 1$$
$$x \geq \dfrac{7}{2} \quad \text{or} \quad x \leq -\dfrac{1}{2}$$

The solution set is $\left\{x \mid x \leq -\dfrac{1}{2} \text{ or } x \geq \dfrac{7}{2}\right\}$.

The correct choice is (**3**).

**19.** Solve the given equation for tan $x$: $\qquad \sqrt{3} \tan x + 1 = 0$

$$\sqrt{3} \tan x = -1$$

$$\tan x = -\frac{1}{\sqrt{3}}$$

Use your calculator to find the reference angle: $\qquad \tan^{-1}\left(-\frac{1}{\sqrt{3}}\right) = -30°$

It is given that $x$ is in the interval $0° \leq x \leq 180°$. Since the value of tan $x$ is negative, the angle terminates in Quadrant II. Hence, $x = 180 - 30 = 150°$.

The correct choice is **(4)**.

**20.** It is given that $\overline{CA} \perp \overline{AB}$, $\overline{ED} \perp \overline{DF}$, $\overline{ED} \parallel \overline{AB}$, $\overline{CE} \cong \overline{BF}$, $\overline{AB} \cong \overline{ED}$, and m$\angle CAB =$ m$\angle FDE = 90$, as shown in the accompanying diagram.

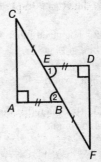

On the basis of the given information, we also know the following:

- $\angle A \cong \angle D$, because they are right angles.
- $\angle 1 \cong \angle 2$, because parallel lines form congruent alternate interior angles.
- $\overline{CB} \cong \overline{FE}$, by adding $\overline{EB}$ to congruent segments $\overline{CE}$ and $\overline{BF}$.
- $\triangle ABC$ and $\triangle DEF$ are right triangles.

These facts would allow us to prove $\triangle ABC \cong \triangle DEF$ using the side-angle-side, angle-angle-side, or hypotenuse-leg theorems. Although we know that two sides of these triangles are congruent, we do not know about the third sides, $\overline{AC}$ and $\overline{DF}$. Hence we could not prove the triangles congruent using the side-side-side theorem (SSS $\cong$ SSS).

The correct choice is **(1)**.

# PART II

**21.** It is given that the function $h(t) = -16t^2 + 45t + 7$ represents the distance, in feet, that the tennis ball is from the ground at any time $t$. The graph of a quadratic equation of the form $y = at^2 + bt + c$ is a parabola. The maximum height of the ball is at the vertex (turning point) of this parabola. This vertex lies on the axis of symmetry, whose equation is $t = -\dfrac{b}{2a}$.

To find the value of $t$ when the ball is at its maximum height, substitute $a = -16$ and $b = 45$:

$$t = -\frac{b}{2a} = -\frac{45}{2(-16)} = 1.40625$$

To the *nearest tenth of a second*, the ball reaches its maximum height at **1.4**.

**22.** The given functions are $f(x) = 2^x - 1$ and $g(x) = x^2 - 1$. To evaluate $f \circ g(3)$, first find $g(3)$: $g(3) = 3^2 - 1 = 8$.

Then $f \circ g(3) = f(g(3)) = f(8) = 2^8 - 1 = \mathbf{255}$.

**23.** It is given that the pressure of air in an automobile tire is inversely proportional to volume. Let $P$ = pressure, and $V$ = volume.

If two variables vary inversely, then their product is a constant. Call this constant $k$:                 $PV = k$

It is given that the pressure is 35 pounds when the volume is 120 cubic inches. Substitute $P = 35$ and $V = 120$ to find the value of $k$:

$$35 \cdot 120 = k$$
$$4{,}200 = k$$

To find the pressure when the volume is 140 cubic inches, substitute $V = 140$:

$$140P = 4{,}200$$
$$P = 30$$

The tire pressure, in pounds, is **30**.

**24.** It is given that the interval of ages between 24.6 and 37.4 years is centered about the mean. Hence the mean age is the midpoint of the interval, or $\dfrac{24.6 + 37.4}{2}$ = 31 years.

It is also given that the ages are normally distributed, and that 95.4% are in the given interval. In a normal distribution, 95.4% of the distribution is within 2 standard deviations of the mean. The distance from the mean to an endpoint of the interval, 37.4 − 31 = 6.4 years, must be 2 standard deviations. Hence, the standard deviation is 6.4 ÷ 2 = 3.2 years.

The mean age is **31** years, and the standard deviation of the data is **3.2** years.

**25.** The given expression is:

$$\frac{9 - x^2}{10x^2 - 28x - 6}$$

To reduce this fraction, first factor the numerator and denominator. The numerator is the difference of two perfect squares. In the denominator 2 is a common factor of all terms. Factor the remaining trinomial into the product of two binomials:

$$\frac{(3 - x)(3 + x)}{2(5x^2 - 14x - 3)}$$

$$\frac{(3 - x)(3 + x)}{2(5x + 1)(x - 3)}$$

Note that $(x - 3) = -(3 - x)$, and reduce the fraction by canceling those factors in the numerator and denominator:

$$\frac{-\cancel{(x - 3)}(3 + x)}{2(5x + 1)\cancel{(x - 3)}}$$

Use the distributive property to simplify the numerator and denominator:

$$\frac{-x - 3}{10x + 2}$$

In simplest form the rational expression is $\dfrac{-x - 3}{10x + 2}$.

**26.** The given expression, $2\displaystyle\sum_{n=1}^{5} (2n - 1)$, represents twice the sum of the terms $(2n - 1)$ for integer values of $n$ from 1 to 5. To evaluate the summation, substitute each of the values $n = 1, 2, 3, 4,$ and 5 into the expression, add the results, and simplify:

$$2\sum_{n=1}^{5} (2n - 1) = 2[(2 \cdot 1 - 1) + (2 \cdot 2 - 1) + (2 \cdot 3 - 1) + (2 \cdot 4 - 1) + (2 \cdot 5 - 1)]$$

$$= 2[1 + 3 + 5 + 7 + 9]$$

$$= 50$$

The value of the given expression is **50**.

# PART III

**27.** It is given that the coordinates of the vertices of quadrilateral *ABCD* are
*A*(−1,−5), *B*(8,2), *C*(11,13), and *D*(2,6), as shown in the accompanying diagram.

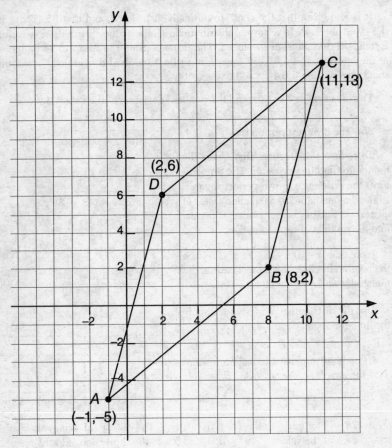

The length of a segment whose endpoints are $(x_1, y_1)$ and $(x_2, y_2)$ can be determined using the distance formula:

$$d = \sqrt{(x_2 - x_1)^2 + (y_2 - y_1)^2}$$

Find the lengths of all four sides of the quadrilateral:

- To find the length of $\overline{AB}$, let $(x_1, y_1) = A(-1,-5)$ and $(x_2, y_2) = B(8,2)$:

$$AB = \sqrt{\left(8 - (-1)\right)^2 + \left(2 - (-5)\right)^2} = \sqrt{9^2 + 7^2} = \sqrt{130}$$

- To find the length of $\overline{BC}$, let $(x_1, y_1) = B(8,2)$ and $(x_2, y_2) = C(11,13)$:

$$BC = \sqrt{\left(11 - 8\right)^2 + \left(13 - 2\right)^2} = \sqrt{3^2 + 11^2} = \sqrt{130}$$

- To find the length of $\overline{DC}$, let $(x_1, y_1) = D(2,6)$ and $(x_2, y_2) = C(11,13)$:

$$DC = \sqrt{\left(11 - 2\right)^2 + \left(13 - 6\right)^2} = \sqrt{9^2 + 7^2} = \sqrt{130}$$

- To find the length of $\overline{AD}$, let $(x_1, y_1) = A(-1,-5)$ and $(x_2, y_2) = D(2,6)$:

$$AD = \sqrt{\left(2 - (-1)\right)^2 + \left(6 - (-5)\right)^2} = \sqrt{3^2 + 11^2} = \sqrt{130}$$

All four sides of quadrilateral $ABCD$ are congruent, because each has a length of $\sqrt{130}$. If all four sides of a quadrilateral are congruent, then the quadrilateral is a rhombus. Hence, $ABCD$ is a **rhombus**.

**28.** It is given that the prices of two stocks over a 12-month period are described by the equations $A(x) = 0.75x^2 - 6x + 20$ and $B(x) = 2.75x + 1.50$.

Use your calculator to make a table of values for the prices of stocks $A$ and $B$ for the interval $0 \le x \le 12$:

| $x$ | $A(x)$ | $B(x)$ |
|-----|--------|--------|
| 0   | 20     | 1.5    |
| 2   | 11     | 7      |
| 4   | 8      | 12.5   |
| 6   | 11     | 18     |
| 8   | 20     | 23.5   |
| 10  | 35     | 29     |
| 12  | 56     | 34.5   |

Plot the points for function $A$, and connect them with a smooth curve in the shape of a parabola. Label this graph $A$, as shown in the accompanying diagram.

Plot the points for function $B$, and connect them with a straight line. Label this graph $B$, as shown in the accompanying diagram.

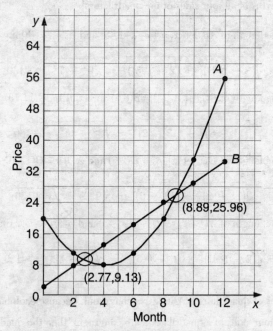

Use your calculator to estimate the coordinates of the points where the two curves intersect. The approximate values are (2.77,9.13) and (8.89,25.96).

The prices, to the *nearest dollar*, when both stock values were the same are **9** and **26**.

**29.** It is given that the path of the male figure skater is represented by the equation $m(x) = 3 \sin \frac{1}{2}x$. The graph is a sine curve with an amplitude of 3 and a period of $4\pi$ units. To graph this curve on the interval $0 \le x \le 4\pi$, start at the origin and draw the first arch upward to reach its maximum height of 3 at $x = \pi$, then downward to cross the x-axis at $x = 2\pi$. Continue downward to the minimum of $-3$ at $x = 3\pi$, then upward to the x-axis at $x = 4\pi$. Label this graph $m$, as shown in the accompanying diagram.

It is also given that the path of the female figure skater is represented by the equation $f(x) = -2 \cos x$. The graph is a cosine curve reflected across the x-axis, having an amplitude of 2 and a period of $2\pi$ units. To graph this curve on the interval $0 \le x \le 4\pi$, start at the y-intercept of $-2$ and draw the curve upward to cross the x-axis at $x = \pi/2$ and reach its maximum height of 2 at $x = \pi$. Complete the first cycle by drawing the curve downward, across the x-axis at $x = 3\pi/2$, and returning to its minimum value of –2 at $x = 2\pi$. Repeat this process to draw the second cycle of the graph between $x = 2\pi$ and $x = 4\pi$. Label this graph f, as shown in the accompanying diagram.

Mark the points where the curves intersect.

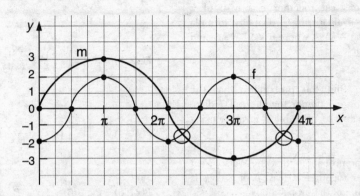

The paths of the skaters intersect **two** times between $x = 0$ and $x = 4\pi$.

**30.** It is given that Sean invests \$10,000 at an annual rate of 5% compounded continuously.

The given formula is:

$$A = Pe^{rt}$$

To find the amount Sean will have after 2 years, substitute $P = 10{,}000$, $e = 2.718$, $r = 0.05$, $t = 2$:

$$A = 10000(2.718)^{0.05(2)}$$

Evaluate using your calculator:

$$\approx 11051.59$$

To the *nearest dollar*, after 2 years Sean will have **11,052**.

To determine how many years it will take for his initial investment to double, substitute $A = 20{,}000$, $P = 10{,}000$, $e = 2.718$, $r = 0.05$, and solve the equation for $t$:

$$20000 = 10000(2.718)^{0.05t}$$

$$2 = (2.718)^{0.05t}$$

$$\log 2 = (0.05t)\log 2.718$$

$$t = \frac{\log 2}{0.05 \log 2.718}$$

Evaluate using your calculator:

$$\approx 13.864$$

To the *nearest year*, it will take **14** for Sean's initial investment to double.

**31.** It is given that, on any given day, the probability that the entire Watson family eats dinner together is $\frac{2}{5}$. To eat together *at least* six times in a week, the Watsons must eat together on 6 or 7 of the 7 days.

To find the probability that the Watsons eat together exactly 6 of 7 days, evaluate the probability formula $_nC_x p^x (1-p)^{n-x}$, where $n = 7$, $x = 6$, and $p = \frac{2}{5}$:

$$P(x = 6) = {}_7C_6 \left(\frac{2}{5}\right)^6 \left(\frac{3}{5}\right)^1 = \frac{1,344}{78,125}$$

To find the probability that the Watson eat dinner together all 7 days, evaluate the probability formula using $x = 7$:

$$P(x = 7) = {}_7C_7 \left(\frac{2}{5}\right)^7 \left(\frac{3}{5}\right)^0 = \frac{128}{78,125}$$

The probability that the Watsons eat dinner together at least 6 of 7 days is the sum of the probabilities:

$$P(x \geq 6) = P(x = 6) + P(x = 7)$$

$$= \frac{1,344}{78,125} + \frac{128}{78,125}$$

$$= \frac{1,472}{78,125}$$

The probability that, during any 7-day period, the Watsons eat dinner together at least six times is $\dfrac{1,472}{78,125}$.

**32.** In the accompanying diagram, $F$ represents the boat's first location, $S$ the second location, and $\overline{TL}$ the lighthouse. It is given that m$\angle F$ = 3, $FS$ = 700 feet, and m$\angle TSL$ = 5, as shown.

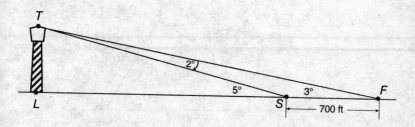

To find the height of the lighthouse, first find the distance from $T$ to $S$ using the Law of Sines. In $\triangle FST$, m$\angle FST$ = 180 – 5 = 175. Since the sum of the angles of any triangle is 180°, m$\angle FTS$ = 180 – 175 – 3 = 2.

Find $ST$ using the Law of Sines and your calculator:

$$\frac{\sin 3}{ST} = \frac{\sin 2}{700}$$

$$ST \sin 2 = 700 \sin 3$$

$$ST = \frac{700 \sin 3}{\sin 2}$$

$$\approx 1049.733$$

Now use the sine ratio in right triangle $SLT$ to find $TL$:

$$\frac{TL}{1049.733} = \sin 5$$

$$TL = 1049.733 \sin 5$$

$$\approx 91.4903$$

To the *nearest tenth of a foot*, the height of the lighthouse is **91.5**.

# PART IV

**33.** It is given that the lengths of the sides of a triangular piece of land are 50, 75, and 100 yards, as shown in the accompanying diagram.

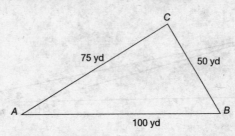

To find the area of the land, first use the Law of Cosines and your calculator to find the measure of ∠A.

$$a^2 = b^2 + c^2 - 2bc \cos A$$

$$50^2 = 75^2 + 100^2 - 2(75)(100) \cos A$$

$$2(75)(100) \cos A = 75^2 + 100^2 - 50^2$$

$$\cos A = \frac{75^2 + 100^2 - 50^2}{2(75)(100)}$$

$$A = 0.875$$

$$A = \cos^{-1}(0.875)$$

$$\approx 28.955°$$

The area, $K$, of a triangle is one-half the product of the lengths of any two sides and the sine of the included angle. Substitute the known measurements, and use your calculator to evaluate the area:

$$K = \frac{1}{2}(75)(100) \sin 28.955 \approx 1,815.4595 \text{ square yards}$$

It is given that the farmer's yearly profit from a crop of strawberries is $1.50 per square yard, or $1.50(1,815.4595) \approx 2,723.19$.

To the *nearest hundred dollars* the farmer can expect a profit of **2,700**.

**34.** *a*   It is given that the equation of circle $V$ is $x^2 + y^2 = 25$. The equation of a circle with radius $r$ and center at the origin is $x^2 + y^2 = r^2$. If $r^2 = 25$, then $r = 5$. Draw a circle with center at the origin and a radius of 5. Label this circle $V$, as shown in the accompanying diagram.

It is also given that the equation of circle $M$ is $(x - 8)^2 + (y + 6)^2 = 4$. In general, the equation of a circle with center at point $(h,k)$ and radius $r$ is $(x - h)^2 + (y - k)^2 = r^2$. The given equation can be written as $(x - 8)^2 + (y - (-6))^2 = 2^2$. Draw a circle with center at point $(8,-6)$ and a radius of 2. Label this circle $M$, as shown in the accompanying diagram.

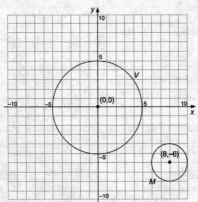

*b*   The probability that a randomly selected point, $(x, y)$, lies in one of the circles is the area of the circle divided by the area of the region. The region such that $-10 \le x \le 10$ and $-10 \le y \le 10$ is a square with sides of length 20; hence its area is 400 square units.

Since the radius of circle $V$ is 5, its area is $\pi \cdot 5^2 = 25\pi$ square units. Hence, the probability that a randomly selected point lies inside circle $V$ is $\dfrac{25\pi}{400}$.

Similarly, the radius of circle $M$ is 2, so its area is $\pi \cdot 2^2 = 4\pi$ square units. Hence, the probability that a randomly selected point lies inside circle $M$ is $\dfrac{4\pi}{400}$.

The probability that a randomly selected point is in either circle $M$ or circle $V$ is the sum of these probabilities:

$$\frac{25\pi}{400} + \frac{4\pi}{400} = \frac{29\pi}{400} \approx 0.22777$$

Therefore, the probability that the randomly selected point is *not* in either circle is $1 - 0.22777 = 0.77223$.

The probability that a randomly selected point, $(x, y)$, lies outside both circles is **0.77223**.

| Topic | Question Numbers | Number of Points | Your Points |
|---|---|---|---|
| 1. Properties of Real Numbers | — | — | |
| 2. Sequences | — | — | |
| 3. Complex Numbers | 4, 15 | 2 + 2 = 4 | |
| 4. Inequalities, Absolute Value | 18 | 2 | |
| 5. Algebraic Expressions, Fractions | 17, 25 | 2 + 2 = 4 | |
| 6. Exponents (zero, negative, rational, scientific notation) | 8 | 2 | |
| 7. Radical Expressions | 5 | 2 | |
| 8. Quadratic Equations (factors, formula, discriminant) | 7 | 2 | |
| 9. Systems of Equations | 28 | 4 | |
| 10. Functions (graphs, domain, range, roots) | 10 | 2 | |
| 11. Inverse Functions, Composition | 22 | 2 | |
| 12. Linear Functions | — | — | |
| 13. Parabolas (max/min, axis of symmetry) | 21 | 2 | |
| 14. Hyperbolas (including inverse variation) | 23 | 2 | |
| 15. Ellipses | 11 | 2 | |
| 16. Exponents and Logarithms | 1, 3, 16, 30 | 2 + 2 + 2 + 4 = 10 | |
| 17. Trig. (circular functions, unit circle, radians) | 2, 19, 29 | 2 + 2 + 4 = 8 | |
| 18. Trig. Equations and Identities | 12 | 2 | |
| 19. Solving Triangles (sin/cos laws, Pythag. thm., ambig. case) | 32, 33 | 4 + 6 = 10 | |
| 20. Coordinate Geom. (slope, distance, midpoint, circle) | 34a | 2 | |
| 21. Transformations, Symmetry | 6, 9, 13, 14 | 2 + 2 + 2 + 2 = 8 | |
| 22. Circle Geometry | — | — | |
| 23. Congruence, Similarity, Proportions | 20 | 2 | |
| 24. Probability (including Bernoulli events) | 31, 34b | 4 + 4 = 8 | |
| 25. Normal Curve | 24 | 2 | |
| 26. Statistics (center, spread, summation notation) | 26 | 2 | |
| 27. Correlation, Modeling, Prediction | — | — | |
| 28. Algebraic Proofs, Word Problems | — | — | |
| 29. Geometric Proofs | — | — | |
| 30. Coordinate Geometry Proofs | 27 | 4 | |

**Total Raw Score =** _____

## HOW TO CONVERT YOUR RAW SCORE
## TO YOUR MATH B REGENTS EXAMINATION SCORE

Below is the conversion chart that must be used to determine your final score on the June 2003 Regents Examination in Math B. To find your final exam score, locate in the column labeled "Raw Score" the total number of points you scored. Then locate in the adjacent column to the right the scaled score that corresponds to your raw score. The scaled score is your final Math B Regents Examination score.

### Regents Examination in Math B—June 2003
### Chart for Converting Total Test Raw Scores to
### Final Examination Scores (Scaled Scores)

| Raw Score | Scaled Score | Raw Score | Scaled Score | Raw Score | Scaled Score |
|---|---|---|---|---|---|
| 88 | 100 | 58 | 75 | 28 | 49 |
| 87 | 99 | 57 | 74 | 27 | 48 |
| 86 | 98 | 56 | 74 | 26 | 47 |
| 85 | 97 | 55 | 73 | 25 | 46 |
| 84 | 96 | 54 | 72 | 24 | 44 |
| 83 | 95 | 53 | 72 | 23 | 43 |
| 82 | 94 | 52 | 71 | 22 | 42 |
| 81 | 93 | 51 | 70 | 21 | 40 |
| 80 | 92 | 50 | 69 | 20 | 39 |
| 79 | 91 | 49 | 69 | 19 | 37 |
| 78 | 90 | 48 | 68 | 18 | 36 |
| 77 | 89 | 47 | 67 | 17 | 34 |
| 76 | 88 | 46 | 66 | 16 | 33 |
| 75 | 87 | 45 | 66 | 15 | 31 |
| 74 | 87 | 44 | 65 | 14 | 29 |
| 73 | 86 | 43 | 64 | 13 | 27 |
| 72 | 85 | 42 | 63 | 12 | 26 |
| 71 | 84 | 41 | 62 | 11 | 24 |
| 70 | 84 | 40 | 61 | 10 | 22 |
| 69 | 83 | 39 | 61 | 9 | 20 |
| 68 | 82 | 38 | 60 | 8 | 18 |
| 67 | 81 | 37 | 59 | 7 | 16 |
| 66 | 81 | 36 | 58 | 6 | 14 |
| 65 | 80 | 35 | 57 | 5 | 12 |
| 64 | 79 | 34 | 56 | 4 | 9 |
| 63 | 78 | 33 | 55 | 3 | 7 |
| 62 | 78 | 32 | 54 | 2 | 5 |
| 61 | 77 | 31 | 53 | 1 | 2 |
| 60 | 76 | 30 | 52 | 0 | 0 |
| 59 | 76 | 29 | 50 | | |

# Examination
# August 2003
## Math B

**FORMULAS**

### Area of Triangle

$$K = \frac{1}{2}ab\sin C$$

### Function of the Sum of Two Angles

$$\sin(A + B) = \sin A \cos B + \cos A \sin B$$
$$\cos(A + B) = \cos A \cos B - \sin A \sin B$$

### Function of the Difference of Two Angles

$$\sin(A - B) = \sin A \cos B - \cos A \sin B$$
$$\cos(A - B) = \cos A \cos B + \sin A \sin B$$

### Law of Sines

$$\frac{a}{\sin A} = \frac{b}{\sin B} = \frac{c}{\sin C}$$

### Law of Cosines

$$a^2 = b^2 + c^2 - 2bc\cos A$$

### Functions of the Double Angle

$$\sin 2A = 2\sin A \cos A$$
$$\cos 2A = \cos^2 A - \sin^2 A$$
$$\cos 2A = 2\cos^2 A - 1$$
$$\cos 2A = 1 - 2\sin^2 A$$

## Functions of the Half Angle

$$\sin \frac{1}{2}A = \pm\sqrt{\frac{1-\cos A}{2}}$$

$$\cos \frac{1}{2}A = \pm\sqrt{\frac{1+\cos A}{2}}$$

## Normal Curve
## Standard Deviation

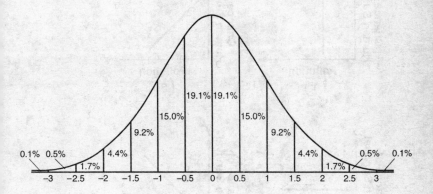

# PART I

Answer all questions in this part. Each correct answer will receive 2 credits. No partial credit will be allowed. For each question, write in the spaces provided the numeral preceding the word or expression that best completes the statement or answers the question.   [40]

1 Which graph does *not* represent a function of x?

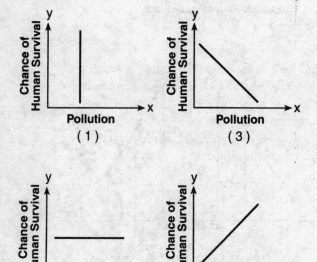

1 ____

2 What is the value of x in the equation $\sqrt{5-2x} = 3i$?

(1) 1                        (3) –2
(2) 7                        (4) 4

2 ____

3 Which graph represents the solution set of $|2x - 1| < 7$?

(1)

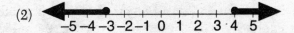

(2)

(3)

(4)

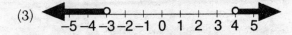

3 ___

4 The strength of a medication over time is represented by the equation $y = 200(1.5)^{-x}$, where $x$ represents the number of hours since the medication was taken and $y$ represents the number of micrograms per millimeter left in the blood. Which graph best represents this relationship?

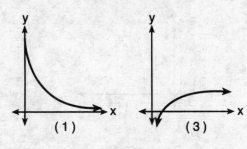

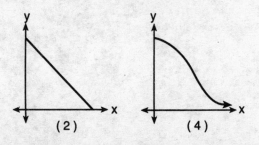

4 ___

5 Written in simplest form, the expression $\dfrac{x^2y^2 - 9}{3 - xy}$ is equivalent to

(1) $-1$

(3) $-(3 + xy)$

(2) $\dfrac{1}{3 + xy}$

(4) $3 + xy$

5 ___

6 Which graph represents data used in a linear regression that produces a correlation coefficient closest to $-1$?

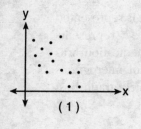

( 1 )

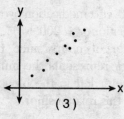

( 3 )

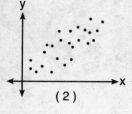

( 2 )

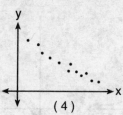

( 4 )

6 ___

7 Which expression is equal to $\dfrac{2+\sqrt{3}}{2-\sqrt{3}}$?

(1) $\dfrac{1-4\sqrt{3}}{7}$          (3) $1-4\sqrt{3}$

(2) $\dfrac{7+4\sqrt{3}}{7}$          (4) $7+4\sqrt{3}$          7 ____

8 Which transformation is *not* an isometry?
   (1) rotation          (3) dilation
   (2) line reflection          (4) translation          8 ____

9 A dog has a 20-foot leash attached to the corner where a garage and a fence meet, as shown in the accompanying diagram. When the dog pulls the leash tight and walks from the fence to the garage, the arc the leash makes is 55.8 feet.

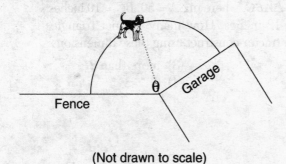

(Not drawn to scale)

What is the measure of angle θ between the garage and the fence, in radians?

(1) 0.36          (3) 3.14
(2) 2.79          (4) 160          9 ____

10 In the accompanying diagram of parallelogram
   *ABCD*, $\overline{DE} \cong \overline{BF}$.

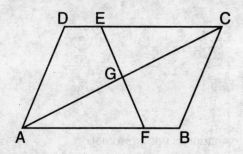

   Triangle *EGC* can be proved congruent to triangle
   *FGA* by

   (1) HL ≅ HL             (3) AAS ≅ AAS
   (2) AAA ≅ AAA           (4) SSA ≅ SSA          10 _____

11 An architect commissions a contractor to produce a
   triangular window. The architect describes the
   window as △*ABC*, where m∠*A* = 50, *BC* = 10 inches,
   and *AB* = 12 inches. How many distinct triangles
   can the contractor construct using these dimensions?

   (1) 1                   (3) more than 2
   (2) 2                   (4) 0                  11 _____

12 The accompanying graph shows the relationship between a person's weight and the distance that the person must sit from the center of a seesaw to make it balanced.

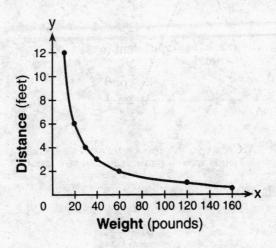

Which equation best represents this graph?

(1) $y = 12x^2$        (3) $y = 2 \log x$

(2) $y = -120x$        (4) $y = \dfrac{120}{x}$        12 _____

13 If f and g are two functions defined by $f(x) = 3x + 5$ and $g(x) = x^2 + 1$, then $g(f(x))$ is

(1) $x^2 + 3x + 6$        (3) $3x^2 + 8$

(2) $9x^2 + 30x + 26$        (4) $9x^2 + 26$        13 _____

14 What is the product of $5 + \sqrt{-36}$ and $1 - \sqrt{-49}$, expressed in simplest $a + bi$ form?

    (1) $-37 + 41i$            (3) $47 + 41i$

    (2) $5 - 71i$             (4) $47 - 29i$       14 _____

15 The expression $\dfrac{2\cos\theta}{\sin2\theta}$ is equivalent to

    (1) $\csc\theta$            (3) $\cot\theta$

    (2) $\sec\theta$            (4) $\sin\theta$       15 _____

16 If $\sin x = \dfrac{12}{13}$, $\cos y = \dfrac{3}{5}$, and $x$ and $y$ are acute angles, the value of $\cos(x - y)$ is

    (1) $\dfrac{21}{65}$            (3) $-\dfrac{14}{65}$

    (2) $\dfrac{63}{65}$            (4) $-\dfrac{33}{65}$       16 _____

17 The amount of ketchup dispensed from a machine at Hamburger Palace is normally distributed with a mean of 0.9 ounce and a standard deviation of 0.1 ounce. If the machine is used 500 times, approximately how many times will it be expected to dispense 1 or more ounces of ketchup?

    (1) 5                (3) 80

    (2) 16               (4) 100       17 _____

18 A commercial artist plans to include an ellipse in a design and wants the length of the horizontal axis to equal 10 and the length of the vertical axis to equal 6. Which equation could represent this ellipse?

(1) $9x^2 + 25y^2 = 225$      (3) $x^2 + y^2 = 100$

(2) $9x^2 - 25y^2 = 225$      (4) $3y = 20x^2$      18 _____

19 A function is defined by the equation $y = \frac{1}{2}x - \frac{3}{2}$.

Which equation defines the inverse of this function?

(1) $y = 2x + 3$      (3) $y = 2x + \frac{3}{2}$

(2) $y = 2x - 3$      (4) $y = 2x - \frac{3}{2}$      19 _____

20 In the equation $ax^2 + 6x - 9 = 0$, imaginary roots will be generated if

(1) $-1 < a < 1$      (3) $a > -1$, only

(2) $a < 1$, only      (4) $a < -1$      20 _____

## PART II

**Answer all questions in this part. Each correct answer will receive 2 credits. Clearly indicate the necessary steps, including appropriate formula substitutions, diagrams, graphs, charts, etc. For all questions in this part, a correct numerical answer with no work shown will receive only 1 credit.**   [12]

21 In height, $h$, in feet, a ball will reach when thrown in the air is a function of time, $t$, in seconds, given by the equation $h(t) = -16t^2 + 30t + 6$. Find, to the *nearest tenth*, the maximum height, in feet, the ball will reach.

22 Find the value of $(x + 2)^0 + (x + 1)^{-\frac{2}{3}}$ when $x = 7$.

23 Express in simplest form: $\dfrac{\dfrac{x}{4} - \dfrac{4}{x}}{1 - \dfrac{4}{x}}$

24 The triangular top of a table has two sides of 14 inches and 16 inches, and the angle between the sides is 30°. Find the area of the tabletop, in square inches.

25 Meteorologists can determine how long a storm lasts by using the function $t(d) = 0.07d^{\frac{3}{2}}$, where $d$ is the diameter of the storm, in miles, and $t$ is the time, in hours. If the storm lasts 4.75 hours, find its diameter, to the *nearest tenth of a mile*.

26 Tom scored 23 points in a basketball game. He attempted 15 field goals and 6 free throws. If each successful field goal is 2 points and each successful free throw is 1 point, is it possible he successfully made all 6 of his free throws? Justify your answer.

## PART III

**Answer all questions in this part. Each correct answer will receive 4 credits. Clearly indicate the necessary steps, including appropriate formula substitutions, diagrams, graphs, charts, etc. For all questions in this part, a correct numerical answer with no work shown will receive only 1 credit.** [24]

27 On the accompanying grid, graph and label $\overline{AB}$, where $A$ is $(0,5)$ and $B$ is $(2,0)$. Under the transformation $r_{x\text{-axis}} \circ r_{y\text{-axis}} (\overline{AB})$, $A$ maps to $A''$, and $B$ maps to $B''$. Graph and label $\overline{A''B''}$. What single transformation would map $\overline{AB}$ to $\overline{A''B''}$?

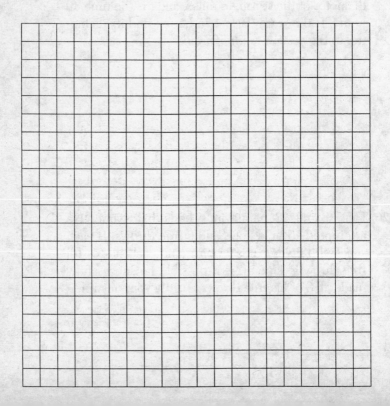

28 Express, in simplest $a + bi$ form, the roots of the equation $x^2 + 5 = 4x$.

29 A ship at sea is 70 miles from one radio transmitter and 130 miles from another. The angle between the signals sent to the ship by the transmitters is 117.4°. Find the distance between the two transmitters, to the *nearest mile*.

30 A student attaches one end of a rope to a wall at a fixed point 3 feet above the ground, as shown in the accompanying diagram, and moves the other end of the rope up and down, producing a wave described by the equation $y = a \sin bx + c$. The range of the rope's height above the ground is between 1 and 5 feet. The period of the wave is $4\pi$. Write the equation that represents this wave.

31 The table below shows the results of an experiment that relates the height at which a ball is dropped, $x$, to the height of its first bounce, $y$.

| Drop Height ($x$) (cm) | Bounce Height ($y$) (cm) |
|:---:|:---:|
| 100 | 26 |
| 90 | 23 |
| 80 | 21 |
| 70 | 18 |
| 60 | 16 |

Find $\bar{x}$, the mean of the drop heights.

Find $\bar{y}$, the mean of the bounce heights.

Find the linear regression equation that best fits the data.

Show that $(\bar{x}, \bar{y})$ is a point on the line of regression. [The use of the accompanying grid is optional.]

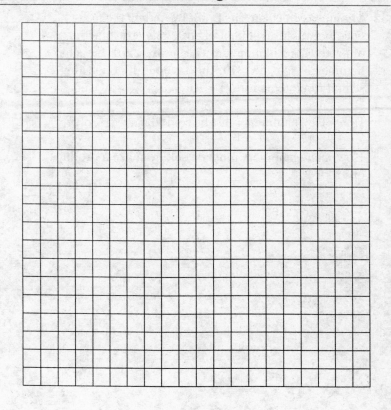

32 A company calculates its profit by finding the difference between revenue and cost. The cost function of producing $x$ hammers is $C(x) = 4x + 170$. If each hammer is sold for $10, the revenue function for selling $x$ hammers if $R(x) = 10x$.

How many hammers must be sold to make a profit?

How many hammers must be sold to make a profit of $100?

## PART IV

Answer all questions in this part. Each correct answer will receive 6 credits. Clearly indicate the necessary steps, including appropriate formula substitutions, diagrams, graphs, charts, etc. For all questions in this part, a correct numerical answer with no work shown will receive only 1 credit.   [12]

33 Given circle $O$ with diameter $\overline{GOAL}$; secants $\overline{HUG}$ and $\overline{HTAM}$ intersect at point $H$; $m\overarc{GM}:m\overarc{ML}:m\overarc{LT}$ = 7:3:2; and chord $\overline{GU} \cong$ chord $\overline{UT}$. Find the ratio of $m\angle UGL$ to $m\angle H$.

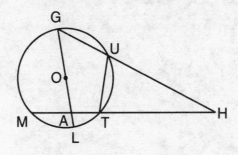

34 When Joe bowls, he can get a strike (knock down all the pins) 60% of the time. How many times more likely is it for Joe to bowl *at least* three strikes out of four tries as it is for him to bowl zero strikes out of four tries? Round your answer to the *nearest whole number.*

# Answers
# August 2003
## Math B

## Answer Key

### PART I

| | | | |
|---|---|---|---|
| 1. 1 | 6. 4 | 11. 2 | 16. 2 |
| 2. 2 | 7. 4 | 12. 4 | 17. 3 |
| 3. 1 | 8. 3 | 13. 2 | 18. 1 |
| 4. 1 | 9. 2 | 14. 4 | 19. 1 |
| 5. 3 | 10. 3 | 15. 1 | 20. 4 |

For Parts II, III, and IV, see **Answers Explained** section for computations and methodology to support the given answers.

### PART II

**21.** 20.1

**22.** $1\frac{1}{4}$

**23.** $\frac{x+4}{4}$

**24.** 56

**25.** 16.6

**26.** No; see **Answers Explained** section.

## PART III

**27.** $A''(0,-5)$, $B''(-2,0)$; $R_{180°}$, $R_{-180°}$, or $r_{(0,0)}$

**28.** $2 \pm i$

**29.** 174

**30.** $y = 2 \sin \dfrac{1}{2} x + 3$ or $y = -2 \sin \dfrac{1}{2} x + 3$

**31.** $\bar{x} = 80$, $\bar{y} = 20.8$, $y = 0.25x + 0.8$; see **Answers Explained** section.

**32.** 29, 45

## PART IV

**33.** $2{:}1, \dfrac{2}{1}$, or an equivalent ratio

**34.** 19

# Answers Explained

## PART I

**1.** In a function each value of $x$ is associated with only one value of $y$. No vertical line will intersect the graph of a function more than once. Since graph (1) is itself a vertical line, an infinite number of values of $y$ are associated with the indicated value of $x$. Hence graph (1) does *not* represent a function of $x$.

The correct choice is **(1)**.

**2.** The equation given is:

$$\sqrt{5 - 2x} = 3i$$

Square both sides of the equation:

$$\left(\sqrt{5 - 2x}\right)^2 = (3i)^2$$

Solve for $x$:

$$5 - 2x = 9i^2$$
$$5 - 2x = -9$$
$$-2x = -14$$
$$x = 7$$

The correct choice is **(2)**.

**3.** The given inequality is:

$$|2x - 1| < 7$$

Rewrite the inequality without the absolute value:

$$-7 < 2x - 1 < 7$$

Solve the inequality:

$$-6 < 2x < 8$$
$$-3 < x < 4$$

The correct choice is **(1)**.

**4.** The equation $y = 200(1.5)^{-x}$ is an exponential function. This function is represented by the graph of a curve with $y$-values that are always positive, decreasing, and approaching the $x$-axis as an asymptote, as seen in graph (1). Check this choice by using your calculator to graph the function.

The correct choice is **(1)**.

**5.** The given expression is:

$$\frac{x^2 y^2 - 9}{3 - xy}$$

To reduce this fraction, first factor the numerator, which is the difference of two perfect squares:

$$\frac{(xy - 3)(xy + 3)}{3 - xy}$$

Note that $(xy - 3) = -(3 - xy)$, and reduce the fraction by canceling those factors in the numerator and denominator:

$$\frac{-\cancel{(xy - 3)}(xy + 3)}{\cancel{3 - xy}}$$

In simplest form, $-(xy + 3)$ and $-(3 + xy)$, given as choice (3), are equivalent.

$$-(xy + 3)$$

The correct choice is **(3)**.

**6.** It is given that a linear regression produces a correlation coefficient near $-1$. Correlation measures the direction and strength of the linear association between two variables. A negative correlation indicates that $y$ generally decreases as $x$ increases; hence the slope of the regression line is negative. A correlation near 1 indicates that the association is very strong; hence the points lie very close to the regression line. Graph (4) shows a strong, negative linear association between $x$ and $y$.

The correct choice is **(4)**.

**7.** The given expression is:

$$\frac{2 + \sqrt{3}}{2 - \sqrt{3}}$$

Multiply the numerator and denominator by the conjugate of the denominator, $2 + \sqrt{3}$:

$$\frac{2 + \sqrt{3}}{2 - \sqrt{3}} \cdot \frac{2 + \sqrt{3}}{2 + \sqrt{3}}$$

Simplify:

$$\frac{\left(2 + \sqrt{3}\right)\left(2 + \sqrt{3}\right)}{\left(2 - \sqrt{3}\right)\left(2 + \sqrt{3}\right)}$$

$$\frac{4 + 4\sqrt{3} + 3}{4 - 3}$$

$$7 + 4\sqrt{3}$$

The correct choice is **(4)**.

**8.** An isometry is a transformation that preserves size. A rotation, a line reflection, and a translation preserve the size of a figure, but a dilation does not.

The correct choice is **(3)**.

**9.** It is given that the length of the leash is 20 feet and the length of the arc the leash makes is 55.8 feet, as shown in the accompanying diagram.

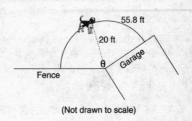

In a circle of radius 20, a central angle of 1 radian intersects an arc 20 feet long.

Hence, this central angle is $\dfrac{55.8}{20} = 2.79$ radians.

(Not drawn to scale)

The correct choice is **(2)**.

**10.** It is given that, in the accompanying diagram of parallelogram $ABCD$, $\overline{DE} \cong \overline{BF}$.

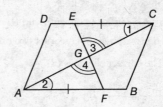

On the basis of the given information, you also know that:

- $\angle 1 \cong \angle 2$, because parallel sides $\overline{CD}$ and $\overline{AB}$ form congruent alternate interior angles.
- $\angle 3 \cong \angle 4$, because vertical angles are congruent.
- $\overline{CE} \cong \overline{AF}$, by subtracting congruent segments $\overline{DE}$ and $\overline{BF}$ from congruent sides $\overline{CD}$ and $\overline{AB}$.

These facts would allow you to prove $\triangle EGC$ congruent to $\triangle FGA$ by AAS $\cong$ AAS.

The correct choice is **(3)**.

**11.** It is given that a contractor is to produce a triangular window, described as $\triangle ABC$, in which m$\angle A$ = 50 and $AB$ = 12 inches, as shown in the accompanying diagram. It is also given that $BC$ = 10 inches.

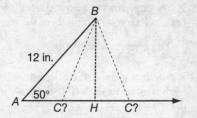

To determine how many distinct triangles the contractor can construct using these dimensions, find the shortest distance from $B$ to the opposite side of the triangle, which is the length of the perpendicular segment $\overline{BH}$. Use the sine function and your calculator:

$$\sin 50 = \frac{BH}{12}$$
$$BH = 12 \sin 50$$
$$\approx 9.19$$

Since $BC$ = 10 is longer than $\overline{BH}$, point $C$ could be at two different locations along $\overrightarrow{AH}$, either between $A$ and $H$ or beyond $H$, as shown in the diagram. Hence the contractor could construct two different triangles using the given dimensions.

The correct choice is **(2)**.

**12.** Each of the points on the graph must satisfy the equation represented. One of the given points has coordinates (20,6), as seen in the accompanying diagram.

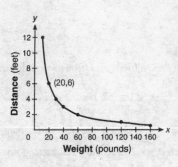

Check $x$ = 20 and $y$ = 6 in each of the equations offered as choices.

- Choice (1): $6 \neq 12 \cdot 20^2$, so $y = 12x^2$ is not the correct equation.
- Choice (2): $6 \neq -120(20)$, so $y = -120x$ is not the correct equation.
- Choice (3): $6 \neq 2 \log 20$, so $y = 2 \log x$ is not the correct equation.

- Choice (4): $6 = \dfrac{120}{20}$, so $y = \dfrac{120}{x}$ *is* the correct equation.

The correct choice is **(4)**.

**13.** The given functions are $f(x) = 3x + 5$ and $g(x) = x^2 + 1$. Then:

$$g(f(x)) = g(3x + 5)$$
$$= (3x + 5)^2 + 1$$
$$= 9x^2 + 30x + 25 + 1$$
$$= 9x^2 + 30x + 26$$

The correct choice is **(2)**.

**14.** The given product is:

$$(5 + \sqrt{-36})(1 - \sqrt{-49})$$

Rewrite each factor as a complex number:

$$(5 + 6i)(1 - 7i)$$

Multiply:

$$5 - 35i + 6i - 42i^2$$

Simplify:

$$5 - 29i + 42$$

$$47 - 29i$$

The correct choice is **(4)**.

**15.** The given expression is:

$$\frac{2 \cos \theta}{\sin 2\theta}$$

To find an equivalent expression, use the identity $\sin 2A = 2 \sin A \cos A$, and then reduce the fraction by canceling common factors of 2 and $\cos \theta$ in the numerator and denominator:

$$\frac{\cancel{2} \cancel{\cos \theta}}{\cancel{2} \sin \theta \cancel{\cos \theta}}$$

$$\frac{1}{\sin \theta}$$

The reciprocal of the sine function is the cosecant function:

$$\csc \theta$$

The correct choice is **(1)**.

**16.** It is given that $x$ is an acute angle and that $\sin x = \dfrac{12}{13}$, as shown in the accompanying diagram. In this 5-12-13 right triangle, $\cos x = \dfrac{5}{13}$.

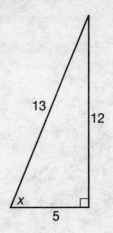

It is also given that $y$ is an acute angle and that $\cos y = \dfrac{3}{5}$, as shown in the accompanying diagram. In this 3-4-5 right triangle, $\sin y = \dfrac{4}{5}$.

Then:

$$\cos (x - y) = \cos x \cos y + \sin x \sin y$$

$$= \frac{5}{13} \cdot \frac{3}{5} \quad + \frac{12}{13} \cdot \frac{4}{5}$$

$$= \frac{63}{65}$$

The correct choice is **(2)**.

**17.** It is given that the amount of ketchup dispensed from a machine is normally distributed with a mean of 0.9 ounce and a standard deviation of 0.1 ounce. One ounce of ketchup is therefore 1 standard deviation above the mean. The accompanying diagram of the normal curve shows that the machine is expected to dispense more than 1 ounce of ketchup approximately 9.2% + 4.4% + 1.7% + 0.5% + 0.1% = 15.9% of the time.

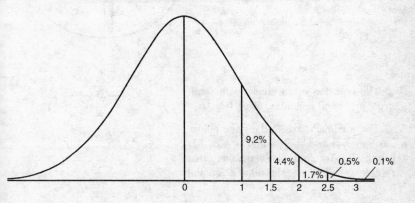

It is also given that the machine is used 500 times. Since 15.9% of 500 is 79.5, the machine can be expected to dispense 1 or more ounces of ketchup about 80 times.

The correct choice is **(3)**.

**18.** It is given that an artist plans to use an ellipse with the lengths of the horizontal and vertical axes equal to 10 and 6, respectively, as shown in the accompanying diagram.

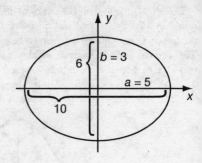

The equation of an ellipse with semimajor axis $a$ and semiminor axis $b$ is:

$$\frac{x^2}{a^2}+\frac{y^2}{b^2}=1$$

The semimajor axis of the given ellipse has length 5 and the semiminor axis has length 3. To find the equation that represents this ellipse, substitute $a = 5$ and $b = 3$:

$$\frac{x^2}{5^2}+\frac{y^2}{3^2}=1$$

$$\frac{x^2}{25}+\frac{y^2}{9}=1$$

To eliminate the denominators, multiply each term by the common denominator $9 \cdot 25$, and simplify:

$$9\cdot\cancel{25}\cdot\frac{x^2}{\cancel{25}}+\cancel{9}\cdot 25\cdot\frac{y^2}{\cancel{9}}=1\cdot 9\cdot 25$$

$$9x^2 + 25y^2 = 225$$

The correct choice is **(1)**.

**19.** The given function is defined as:

$$y = \frac{1}{2}x - \frac{3}{2}$$

To find the inverse, first interchange $x$ and $y$:

$$x = \frac{1}{2}y - \frac{3}{2}$$

Solve the new equation for $y$:

$$x + \frac{3}{2} = \frac{1}{2}y$$

$$2x + 3 = y$$

The correct choice is **(1)**.

**20.** The given equation is:

$$ax^2 + 6x - 9 = 0$$

This equation is in the standard form $ax^2 + bx + c = 0$. Imaginary roots will be generated if the discriminant, $b^2 - 4ac$, is negative.

$$b^2 - 4ac < 0$$

Substitute $b = 6$ and $c = -9$:

$$6^2 - 4a(-9) < 0$$

Solve for $a$:

$$36 + 36a < 0$$

$$36a < -36$$

$$a < -1$$

The correct choice is **(4)**.

# PART II

**21.** It is given that $h$ represents the ball's height, in feet, above the ground $t$ seconds after it is thrown. It is also given that $h(t) = -16t^2 + 30t + 6$, which is a quadratic equation.

The graph of a quadratic equation of the form $y = at^2 + bt + c$ is a parabola. The ball reaches its maximum height at the vertex (turning point) of this parabola. The vertex lies on the axis of symmetry, whose equation is $t = -\dfrac{b}{2a}$. To find the value of $t$ at the turning point, substitute $a = -16$ and $b = 30$:

$$t = -\frac{b}{2a} = -\frac{30}{2(-16)} = 0.9375 \text{ sec}$$

To find the maximum height, $h$, of the ball, substitute $t = 0.9375$ in the given equation:

$$h(0.9375) = -16(0.9375)^2 + 30(0.9375) + 6 = 20.625 \text{ ft}$$

To the *nearest tenth*, the maximum height, in feet, that the ball will reach is **20.1 ft**.

**22.** The given expression is $\qquad\qquad (x + 2)^0 + (x + 1)^{-\frac{2}{3}}$

Substitute $x = 7$: $\qquad\qquad (7 + 2)^0 + (7 + 1)^{-\frac{2}{3}}$

Simplify: $\qquad\qquad\qquad\qquad 9^0 + 8^{-\frac{2}{3}}$

$$1 + 2^{-2}$$

$$1 + \frac{1}{4}$$

The value of the expression is $1\dfrac{1}{4}$.

**23.** The given expression is:

$$\dfrac{\dfrac{x}{4}-\dfrac{4}{x}}{1-\dfrac{4}{x}}$$

To simplify this complex fraction, multiply the numerator and denominator by the common denominator of the fraction terms, $4x$:

$$\dfrac{4x\left(\dfrac{x}{4}-\dfrac{4}{x}\right)}{4x\left(1-\dfrac{4}{x}\right)}$$

Distribute, and simplify:

$$\dfrac{4x\left(\dfrac{x}{4}\right)-4x\left(\dfrac{4}{x}\right)}{4x(1)-4x\left(\dfrac{4}{x}\right)}$$

$$\dfrac{x^2-16}{4x-16}$$

Factor the numerator and denominator, and cancel the common factor, $(x-4)$:

$$\dfrac{(x+4)\cancel{(x-4)}}{4\cancel{(x-4)}}$$

$$\dfrac{x+4}{4}$$

In simplest form, the given expression is equivalent to $\dfrac{x+4}{4}$.

**24.** It is given that two sides of a triangular tabletop measure 14 inches and 16 inches, and the included angle measures 30°, as shown in the accompanying diagram.

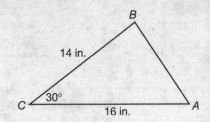

The area, $K$, of a triangle is one-half the product of any two sides and the sine of the included angle:

$$K = \frac{1}{2}\,ab \sin C$$

$$= \frac{1}{2}(14)(16) \sin 30$$

Substitute the given measurements, and use your calculator to evaluate the area:

$$= 56$$

The area of the tabletop, in square inches, is **56**.

**25.** The given function, where $d$ = diameter, in miles, of the storm, and $t$ = time, in hours, is:

$$t(d) = 0.07d^{\frac{3}{2}}$$

Since it is given that the storm lasts 4.75 hours:

$$4.75 = 0.07d^{\frac{3}{2}}$$

To find the diameter of the storm, solve the equation for $d$, evaluating the result with your calculator:

$$\frac{4.75}{0.07} = d^{\frac{3}{2}}$$

$$\left(\frac{4.75}{0.07}\right)^{\frac{2}{3}} = d$$

$$16.6366 \approx d$$

To the *nearest tenth of a mile*, the diameter of the storm is **16.6**.

**26.** No. Suppose Tom successfully made all 6 of his free throws, for 6 points. Then, since it is given that he scored a total of 23 points, he would need to have scored 23 – 6 = 17 points on field goals. It is impossible to score 17 points, an odd number, on field goals, because field goals are worth 2 points each. Hence, it is not possible that Tom made all 6 of his free throws.

# PART III

**27.** The accompanying diagram shows the graph of the given segment $\overline{AB}$.

To graph $\overline{A''B''}$, the image of $\overline{AB}$ under the transformation $r_{x\text{-axis}} \circ r_{y\text{-axis}}$, first reflect the graph of $\overline{AB}$ across the $y$-axis. The image $\overline{A'B'}$, shown as a dotted segment in the diagram, has endpoints at $A'(0,5)$ and $B'(-2,0)$.

Now reflect $\overline{A'B'}$ across the $x$-axis. The final image is $\overline{A''B''}$, the segment with endpoints at $\mathbf{A''(0,-5)}$ and $\mathbf{B''(-2,0)}$, as seen in the diagram.

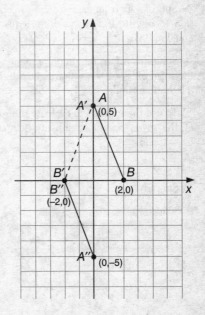

The single transformation $\mathbf{r_{(0,0)}}$, a reflection through the origin, would map $\overline{AB}$ to $\overline{A''B''}$. Note that a rotation of 180°, either clockwise, $\mathbf{R_{-180°}}$, or counterclockwise, $\mathbf{R_{180°}}$, would also accomplish the desired mapping.

**28.** The given equation is:

$$x^2 + 5 = 4x$$

Subtract $4x$ from each side of the equation to rewrite it as a quadratic equation equal to 0:

$$x^2 - 4x + 5 = 0$$

Solutions of a quadratic equation in the form $ax^2 + bx + c = 0$ are given by the quadratic formula:

$$x = \frac{-b \pm \sqrt{b^2 - 4ac}}{2a}$$

To use the formula to evaluate the roots of the given equation, substitute $a = 1$, $b = -4$, and $c = 5$:

$$= \frac{-(-4) \pm \sqrt{(-4)^2 - 4 \cdot 1 \cdot 5}}{2 \cdot 1}$$

Simplify:

$$= \frac{4 \pm \sqrt{16 - 20}}{2}$$

$$= \frac{4 \pm \sqrt{-4}}{2}$$

$$= \frac{4 \pm 2i}{2}$$

$$= 2 \pm i$$

In simplest $a + bi$ form, the roots are **2 + $i$** and **2 − $i$**.

**29.** It is given that a ship is 70 miles from one radio transmitter and 130 miles from another, and also that the angle between the signals sent to the ship by the transmitters is 117.4°, as shown in the accompanying diagram.

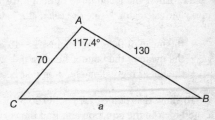

Use the Law of Cosines and your calculator to estimate $a$, the distance between the two transmitters:

$$a^2 = b^2 + c^2 - 2bc \cos A$$

$$= 70^2 + 130^2 - 2(70)(130) \cos 117.4$$

$$\approx 30175.636$$

$$a \approx 173.71$$

To the *nearest mile*, the distance between the transmitters is **174**.

**30.** An equation of the given form, $y = a \sin bx + c$, describes a wave, where $a$ determines the amplitude of the wave, $b$ determines the period of the wave, and $c$ represents a vertical translation of the wave.

It is also given that the range of the rope's height above the ground is between 1 and 5 feet. Hence the amplitude of the wave is $\frac{1}{2}(5-1) = 2$, so $a = 2$ or $-2$.

A sine wave with amplitude 2 would reach a maximum height of 2 feet. Since the given wave reaches a maximum height of 5 feet, it must be translated upward 3 feet. Hence $c = 3$.

A wave described by any equation of the form $y = \sin bx$ completes one cycle when $bx = 2\pi$. Since the period of the given wave is $4\pi$, a cycle is complete when $x = 4\pi$. Solving $b \cdot 4\pi = 2\pi$ yields $b = \frac{1}{2}$.

The equation that represents the wave is $y = 2 \sin \frac{1}{2}x + 3$ or $y = -2 \sin \frac{1}{2}x + 3$.

**31.** Enter the given table of data for drop heights $(x)$ and bounce heights $(y)$ into your calculator. Using the calculator's statistics functions, find:

- the mean of the drop heights: $\overline{x} = \mathbf{80\ cm}$
- the mean of the bounce heights: $\overline{y} = \mathbf{20.8\ cm}$
- the linear regression equation that best fits the data: $\boldsymbol{y = 0.25x + 0.8}$

To show that point $(\overline{x}, \overline{y}) = (80, 20.8)$ lies on the line of regression, substitute $\overline{x} = 80$ for $x$ in the equation and find the corresponding value of $y$:

$$y = 0.25(80) + 0.8 = 20.8 = \overline{y}$$

Point $(\overline{x}, \overline{y})$ lies on the line of regression.

**32.** It is given that the cost function of producing $x$ hammers is $C(x) = 4x + 170$ and that the revenue function for selling $x$ hammers is $R(x) = 10x$.

To make a profit selling hammers, the company's revenue must be greater than its cost. Solve the inequality for $x$:

$$R(x) > C(x)$$

$$10x > 4x + 170$$

$$6x > 170$$

$$x > 28.\overline{3}$$

Since the company must sell a whole number of hammers, **29** hammers must be sold to make a profit.

Profit is the difference between revenue and cost. To determine how many hammers must be sold to make a profit of \$100, let the difference between revenue and cost equal 100, and solve the resulting equation for $x$:

$$R(x) - C(x) = 100$$

$$10x - (4x + 170) = 100$$

$$10x - 4x - 170 = 100$$

$$6x = 270$$

$$x = 45$$

The company must sell **45** hammers to make a profit of \$100.

# PART IV

**33.** It is given that, in the accompanying diagram of circle $O$, $\text{m}\widehat{GM}:\text{m}\widehat{ML}:\text{m}\widehat{LT}$ = 7:3:2. Let $\text{m}\widehat{GM}$ = 7$x$, $\text{m}\widehat{ML}$ = 3$x$, and $\text{m}\widehat{LT}$ = 2$x$.

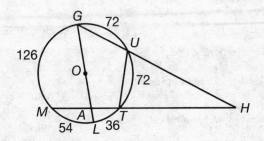

It is also given that $\overline{GOAL}$ is a diameter; hence $\widehat{GML}$ is a semicircle with measure 180°. Write an equation to find the value of $x$:

$$\text{m}\widehat{GM} + \text{m}\widehat{ML} = 180$$

Substitute 7$x$ for $\text{m}\widehat{GM}$ and 3$x$ for $\text{m}\widehat{ML}$:

$$7x + 3x = 180$$
$$10x = 180$$
$$x = 18$$

Since $x$ = 18, then $\text{m}\widehat{GM}$ = 7(18) = 126, $\text{m}\widehat{ML}$ = 3(18) = 54, and $\text{m}\widehat{LT}$ = 2(18) = 36.

$\widehat{GUTL}$ is also a semicircle; hence, $\text{m}\widehat{GUT}$ = 180 − 36 = 144. It is given that $\overline{GU} \cong \overline{UT}$. In a circle, congruent chords have congruent arcs, so $\widehat{GU} \cong \widehat{UT}$. Hence, $\text{m}\widehat{GU} = \text{m}\widehat{UT} = \frac{1}{2}(144) = 72$.

In a circle, the measure of an inscribed angle is one-half the measure of its intercepted arc. Hence, $\text{m}\angle UGL = \frac{1}{2}\text{m}\widehat{LU} = \frac{1}{2}(36 + 72) = 54$.

When two secants are drawn to a circle from the same external point, the measure of the angle formed is one-half the difference of the measures of the intercepted arcs. Hence, $\text{m}\angle H = \frac{1}{2}(\text{m}\widehat{GM} - \text{m}\widehat{TU}) = \frac{1}{2}(126 - 72) = 27$.

Therefore, the ratio of the measures of the angles, $\text{m}\angle UGL : \text{m}\angle H$, is 54:27 = 2:1.
The ratio of $\text{m}\angle UGL$ to $\text{m}\angle H$ is **2:1** or $\dfrac{\mathbf{2}}{\mathbf{1}}$.

**34.** It is given that Joe can get a strike 60% of the time. To bowl *at least* three strikes out of four tries, Joe must bowl strikes on three or four of the four tries.

To find the probability that Joe bowls a strike on exactly three out of four tries, evaluate the probability formula $_nC_x p^x(1-p)^{n-x}$, where $n = 4$, $x = 3$, and $p = 0.6$:

$$P(x = 3) = {_4}C_3(0.6)^3(0.4)^1 = 0.3456$$

To find the probability that Joe bowls a strike on all four tries, evaluate the probability formula using $x = 4$:

$$P(x = 4) = {_4}C_4(0.6)^4(0.4)^0 = 0.1296$$

The probability that Joe bowls at least three strikes is the sum of the probabilities:

$$P(x \geq 3) = P(x = 3) + P(x = 4)$$
$$= 0.3456 + 0.1296$$
$$= 0.4752$$

To find the probability that Joe bowls zero strikes in four tries, evaluate the probability formula using $x = 0$:

$$P(x = 0) = {_4}C_0(0.6)^0(0.4)^4 = 0.0256$$

To determine how many times more likely the first outcome is than the second, divide the probabilities:

$$\frac{0.4752}{0.0256} = 18.5625$$

To the *nearest whole number,* it is **19** times more likely for Joe to bowl *at least* three strikes out of four tries as it is for him to bowl zero strikes out of four tries.

| Topic | Question Numbers | Number of Points | Your Points |
|---|---|---|---|
| 1. Properties of Real Numbers | — | — | |
| 2. Sequences | — | — | |
| 3. Complex Numbers | 2, 14 | 2 + 2 = 4 | |
| 4. Inequalities, Absolute Value | 3 | 2 | |
| 5. Algebraic Expressions, Fractions | 5, 23 | 2 + 2 = 4 | |
| 6. Exponents (zero, negative, rational, scientific notation) | 22 | 2 | |
| 7. Radical Expressions | 7 | 2 | |
| 8. Quadratic Equations (factors, formula, discriminant) | 20, 28 | 2 + 4 = 6 | |
| 9. Systems of Equations | 32 | 4 | |
| 10. Functions (graphs, domain, range, roots) | 1 | 2 | |
| 11. Inverse Functions, Composition | 13, 19 | 2 + 2 = 4 | |
| 12. Linear Functions | — | — | |
| 13. Parabolas (max/min, axis of symmetry) | 21 | 2 | |
| 14. Hyperbolas (including inverse variation) | 12 | 2 | |
| 15. Ellipses | 18 | 2 | |
| 16. Exponents and Logarithms | 4, 25 | 2 + 2 = 4 | |
| 17. Trig. (circular functions, unit circle, radians) | 9, 30 | 2 + 4 = 6 | |
| 18. Trig. Equations and Identities | 15, 16 | 2 + 2 = 4 | |
| 19. Solving Triangles (sin/cos laws, Pythag. thm., ambig. case) | 11, 24, 29 | 2 + 2 + 4 = 8 | |
| 20. Coordinate Geom. (slope, distance, midpoint, circle) | — | — | |
| 21. Transformations, Symmetry | 8, 27 | 2 + 4 = 6 | |
| 22. Circle Geometry | 33 | 6 | |
| 23. Congruence, Similarity, Proportions | 10 | 2 | |
| 24. Probability (including Bernoulli events) | 34 | 6 | |
| 25. Normal Curve | 17 | 2 | |
| 26. Statistics (center, spread, summation notation) | — | — | |
| 27. Correlation, Modeling, Prediction | 6, 31 | 2 + 4 = 6 | |
| 28. Algebraic Proofs, Word Problems | 26 | 2 | |
| 29. Geometric Proofs | — | — | |
| 30. Coordinate Geometry Proofs | — | — | |

Total Raw Score = _____

## HOW TO CONVERT YOUR RAW SCORE
## TO YOUR MATH B REGENTS EXAMINATION SCORE

Below is the conversion chart that must be used to determine your final score on the August 2003 Regents Examination in Math B. To find your final exam score, locate in the column labeled "Raw Score" the total number of points you scored. Then locate in the adjacent column to the right the scaled score that corresponds to your raw score. The scaled score is your final Math B Regents Examination score.

### Regents Examination in Math B—August 2003
### Chart for Converting Total Test Raw Scores to
### Final Examination Scores (Scaled Scores)

| Raw Score | Scaled Score | Raw Score | Scaled Score | Raw Score | Scaled Score |
|-----------|--------------|-----------|--------------|-----------|--------------|
| 88 | 100 | 58 | 76 | 28 | 41 |
| 87 | 99 | 57 | 75 | 27 | 39 |
| 86 | 99 | 56 | 74 | 26 | 38 |
| 85 | 98 | 55 | 73 | 25 | 37 |
| 84 | 97 | 54 | 72 | 24 | 35 |
| 83 | 97 | 53 | 70 | 23 | 34 |
| 82 | 96 | 52 | 69 | 22 | 33 |
| 81 | 95 | 51 | 68 | 21 | 31 |
| 80 | 95 | 50 | 67 | 20 | 30 |
| 79 | 94 | 49 | 66 | 19 | 28 |
| 78 | 93 | 48 | 65 | 18 | 27 |
| 77 | 92 | 47 | 64 | 17 | 26 |
| 76 | 92 | 46 | 63 | 16 | 24 |
| 75 | 91 | 45 | 62 | 15 | 23 |
| 74 | 90 | 44 | 61 | 14 | 21 |
| 73 | 89 | 43 | 59 | 13 | 20 |
| 72 | 88 | 42 | 58 | 12 | 18 |
| 71 | 88 | 41 | 57 | 11 | 17 |
| 70 | 87 | 40 | 56 | 10 | 15 |
| 69 | 86 | 39 | 55 | 9 | 14 |
| 68 | 85 | 38 | 53 | 8 | 12 |
| 67 | 84 | 37 | 52 | 7 | 11 |
| 66 | 83 | 36 | 51 | 6 | 9 |
| 65 | 82 | 35 | 50 | 5 | 8 |
| 64 | 81 | 34 | 49 | 4 | 6 |
| 63 | 80 | 33 | 47 | 3 | 5 |
| 62 | 79 | 32 | 46 | 2 | 3 |
| 61 | 79 | 31 | 45 | 1 | 2 |
| 60 | 78 | 30 | 43 | 0 | 0 |
| 59 | 77 | 29 | 42 | | |

# Examination June 2004
## Math B

**FORMULAS**

### Area of Triangle

$$K = \frac{1}{2} ab \sin C$$

### Function of the Sum of Two Angles

$$\sin (A + B) = \sin A \cos B + \cos A \sin B$$
$$\cos (A + B) = \cos A \cos B - \sin A \sin B$$

### Function of the Difference of Two Angles

$$\sin (A - B) = \sin A \cos B - \cos A \sin B$$
$$\cos (A - B) = \cos A \cos B + \sin A \sin B$$

### Law of Sines

$$\frac{a}{\sin A} = \frac{b}{\sin B} = \frac{c}{\sin C}$$

### Law of Cosines

$$a^2 = b^2 + c^2 - 2bc \cos A$$

### Functions of the Double Angle

$$\sin 2A = 2 \sin A \cos A$$
$$\cos 2A = \cos^2 A - \sin^2 A$$
$$\cos 2A = 2 \cos^2 A - 1$$
$$\cos 2A = 1 - 2 \sin^2 A$$

## Functions of the Half Angle

$$\sin \frac{1}{2} A = \pm\sqrt{\frac{1 - \cos A}{2}}$$

$$\cos \frac{1}{2} A = \pm\sqrt{\frac{1 + \cos A}{2}}$$

## Normal Curve
## Standard Deviation

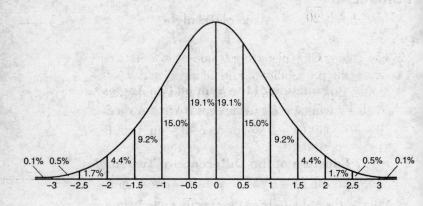

# PART I

Answer all questions in this part. Each correct answer will receive 2 credits. No partial credit will be allowed. For each question, write in the spaces provided the numeral preceding the word or expression that best completes the statement or answers the question. [40]

1 What is the sum of $2 - \sqrt{-4}$ and $-3 + \sqrt{-16}$ expressed in $a + bi$ form?

  (1) $-1 + 2i$              (3) $-1 + 12i$

  (2) $-1 + i\sqrt{20}$          (4) $-14 + i$          1 _____

2 The Hiking Club plans to go camping in a State park where the probability of rain on any give day is 0.7. Which expression can be used to find the probability that it will rain on *exactly* three of the seven days they are there?

  (1) $_7C_3(0.7)^3(0.3)^4$        (3) $_4C_3(0.7)^3(0.7)^4$

  (2) $_7C_3(0.3)^3(0.7)^4$        (4) $_4C_3(0.4)^4(0.3)^3$      2 _____

3 What is the amplitude of the function $y = \dfrac{2}{3} \sin 4x$?

  (1) $\dfrac{\pi}{2}$                   (3) $3\pi$

  (2) $\dfrac{2}{3}$                   (4) $4$                3 _____

4 Which quadratic function is shown in the accompanying graph?

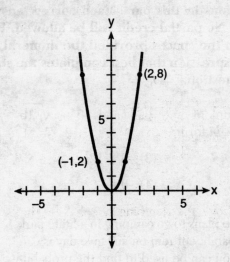

(1) $y = -2x^2$        (3) $y = -\dfrac{1}{2}x^2$

(2) $y = 2x^2$        (4) $y = \dfrac{1}{2}x^2$      4 ____

5 In the accompanying graph, the shaded region represents set $A$ of all points $(x,y)$ such that $x^2 + y^2 \leq 1$. The transformation $T$ maps point $(x,y)$ to point $(2x, 4y)$.

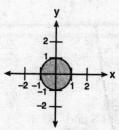

Which graph shows the mapping of set $A$ by the transformation $T$?

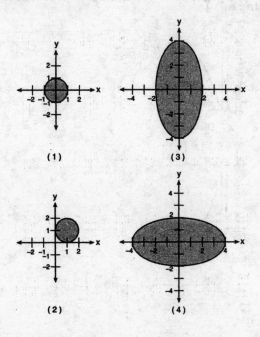

5 _____

6 If $f(x) = 4x^0 + (4x)^{-1}$, what is the value of $f(4)$?

(1) $-12$                 (3) $1\frac{1}{16}$

(2) $0$                   (4) $4\frac{1}{16}$      6 \_\_\_\_

7 What is the domain of the function $f(x) = \dfrac{2x^2}{x^2-9}$?

(1) all real numbers except 0
(2) all real numbers except 3
(3) all real numbers except 3 and $-3$
(4) all real numbers      7 \_\_\_\_

8 Which graph represents an inverse variation between stream velocity and the distance from the center of the stream?

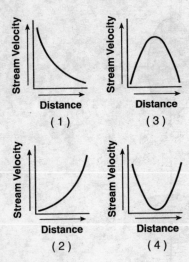

(1)            (3)

(2)            (4)      8 \_\_\_\_

9 If $\log_b x = y$, then $x$ equals

(1) $y \cdot b$            (3) $y^b$

(2) $\dfrac{y}{b}$            (4) $b^y$     9 \_\_\_\_

10 The expression $i^0 \cdot i^1 \cdot i^2 \cdot i^3 \cdot i^4$ is equal to

(1) 1            (3) $i$

(2) $-1$          (4) $-i$     10 \_\_\_\_

11 Which equation models the data in the accompanying table?

| Time in hours, $x$ | 0 | 1 | 2 | 3 | 4 | 5 | 6 |
|---|---|---|---|---|---|---|---|
| Population, $y$ | 5 | 10 | 20 | 40 | 80 | 160 | 320 |

(1) $y = 2x + 5$            (3) $y = 2x$

(2) $y = 2^x$            (4) $y = 5(2^x)$     11 \_\_\_\_

12 The amount of juice dispensed from a machine is normally distributed with a mean of 10.50 ounces and a standard deviation of 0.75 ounce. Which interval represents the amount of juice dispensed about 68.2% of the time?

(1) 9.00–12.00            (3) 9.75–11.25

(2) 9.75–10.50            (4) 10.50–11.25     12 \_\_\_\_

13 If $\theta$ is an acute angle such that $\sin \theta = \dfrac{5}{13}$, what is the value of $\sin 2\theta$?

(1) $\dfrac{12}{13}$            (3) $\dfrac{60}{169}$

(2) $\dfrac{10}{26}$            (4) $\dfrac{120}{169}$     13 \_\_\_\_

14 Which function is symmetrical with respect to the origin?

(1) $y = \sqrt{x+5}$

(3) $y = -\dfrac{5}{x}$

(2) $y = |5 - x|$

(4) $y = 5^x$

14 \_\_\_\_

15 The expression $\dfrac{\dfrac{1}{x} + \dfrac{1}{y}}{\dfrac{1}{x^2} - \dfrac{1}{y^2}}$ is equivalent to

(1) $\dfrac{xy}{y-x}$

(3) $\dfrac{y-x}{xy}$

(2) $\dfrac{xy}{x-y}$

(4) $y - x$

15 \_\_\_\_

16 Sam is designing a triangular piece for a metal sculpture. He tells Martha that two of the sides of the piece are 40 inches and 15 inches, and the angle opposite the 40-inch side measures 120°. Martha decides to sketch the piece that Sam described. How many different triangles can she sketch that match Sam's description?

(1) 1

(3) 3

(2) 2

(4) 0

16 \_\_\_\_

17 If $f(x) = x + 1$ and $g(x) = x^2 - 1$, the expression $(g \circ f)(x)$ equals 0 when $x$ is equal to

(1) 1 and –1

(3) –2, only

(2) 0, only

(4) 0 and –2

17 \_\_\_\_

18 If θ is a positive acute angle and sin θ = $a$, which expression represents cos θ in terms of $a$?

(1) $\sqrt{a}$

(3) $\dfrac{1}{\sqrt{a}}$

(2) $\sqrt{1-a^2}$

(4) $\dfrac{1}{\sqrt{1-a^2}}$   18 ___

19 The expression $\sqrt[4]{16a^6b^4}$ is equivalent to

(1) $2a^2b$

(3) $4a^2b$

(2) $2a^{\frac{3}{2}}b$

(4) $4a^{\frac{3}{2}}b$   19 ___

20 In the accompanying diagram, $\overline{HK}$ bisects $\overline{IL}$ and ∠H ≅ ∠K.

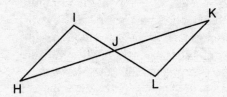

What is the most direct method of proof that could be used to prove $\triangle HIJ \cong \triangle KLJ$?

(1) HL ≅ HL

(3) AAS ≅ AAS

(2) SAS ≅ SAS

(4) ASA ≅ ASA   20 ___

## PART II

Answer all questions in this part. Each correct answer will receive 2 credits. Clearly indicate the necessary steps, including appropriate formula substitutions, diagrams, graphs, charts, etc. For all questions in this part, a correct numerical answer with no work shown will receive only 1 credit. [12]

21 The projected total annual profits, in dollars, for the Nutyme Clothing Company from 2002 to 2004 can be approximated by the model $\sum_{n=0}^{2}(13,567n+294)$, where $n$ is the year and $n = 0$ represents 2002. Use this model to find the company's projected total annual profits, in dollars, for the period 2002 to 2004.

22 Solve algebraically for $x$: $27^{2x+1} = 9^{4x}$

23 Find all values of $k$ such that the equation $3x^2 - 2x + k = 0$ has imaginary roots.

24  In the accompanying diagram of square $ABCD$, $F$ is
the midpoint of $\overline{AB}$, $G$ is the midpoint of $\overline{BC}$, $H$
is the midpoint of $\overline{CD}$, and $E$ is the midpoint of $\overline{DA}$.

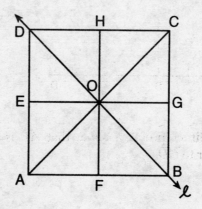

Find the image of $\triangle EOA$ after it is reflected in line $\ell$.
Is this isometry direct or opposite? Explain your
answer.

25  Given:  $\triangle ABT$, $\overline{CBTD}$, and $\overline{AB} \perp \overline{CD}$

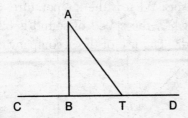

Write an indirect proof to show that $\overline{AT}$ is *not* perpendicular to $\overline{CD}$.

26 The equation $V = 20\sqrt{C + 273}$ relates speed of sound, $V$, in meters per second, to air temperature, $C$, in degrees Celsius. What is the temperature, in degrees Celsius, when the speed of sound is 320 meters per second? [The use of the accompanying grid is optional.]

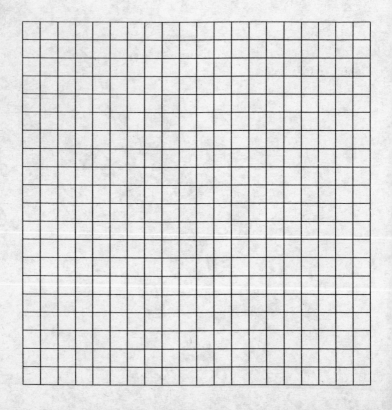

## PART III

Answer all questions in this part. Each correct answer will receive 4 credits. Clearly indicate the necessary steps, including appropriate formula substitutions, diagrams, graphs, charts, etc. For all questions in this part, a correct numerical answer with no work shown will receive only 1 credit.  [24]

27  Navigators aboard ships and airplanes use nautical miles to measure distance. The length of a nautical mile varies with latitude. The length of a nautical mile, $L$, in feet, on the latitude line $\theta$ is given by the formula $L = 6{,}077 - 31 \cos 2\theta$.

Find, to the *nearest degree*, the angle $\theta$, $0° \leq \theta \leq 90°$, at which the length of a nautical mile is approximately 6,076 feet.

28 Two equal forces act on a body at an angle of 80°. If the resultant force is 100 newtons, find the value of one of the two equal forces, to the *nearest hundredth of a newton.*

29 Solve for $x$ and express your answer in simplest radical form:

$$\frac{4}{x} - \frac{3}{x+1} = 7$$

30 A baseball player throws a ball from the outfield toward home plate. The ball's height above the ground is modeled by the equation $y = -16x^2 + 48x + 6$, where $y$ represents height, in feet, and $x$ represents time, in seconds. The ball is initially thrown from a height of 6 feet.

How many seconds after the ball is thrown will it again be 6 feet above the ground?

What is the maximum height, in feet, that the ball reaches? [The use of the accompanying grid is optional.]

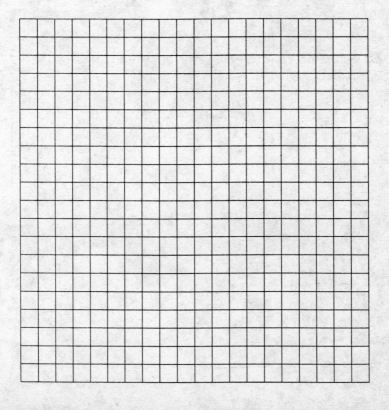

31 An archaeologist can determine the approximate age of certain ancient specimens by measuring the amount of carbon-14, a radioactive substance, contained in the specimen. The formula used to determine the age of a specimen is $A = A_0 2^{\frac{-t}{5760}}$, where $A$ is the amount of carbon-14 that a specimen contains, $A_0$ is the original amount of carbon-14, $t$ is time, in years, and 5760 is the half-life of carbon-14.

A specimen that originally contained 120 milligrams of carbon-14 now contains 100 milligrams of this substance. What is the age of the specimen, to the *nearest hundred years*?

32 Mrs. Ramírez is a real estate broker. Last month, the sale prices of homes in her area approximated a normal distribution with a mean of $150,000 and a standard deviation of $25,000.

A house had a sale price of $175,000. What is the percentile rank of its sale price, to the *nearest whole number*? Explain what that percentile means.

Mrs. Ramírez told a customer that most of the houses sold last month had selling prices between $125,000 and $175,000. Explain why she is correct.

## PART IV

Answer all questions in this part. Each correct answer will receive 6 credits. Clearly indicate the necessary steps, including appropriate formula substitutions, diagrams, graphs, charts, etc. For all questions in this part, a correct numerical answer with no work shown will receive only 1 credit. [12]

33 The accompanying diagram shows a circular machine part that has rods $\overline{PT}$ and $\overline{PAR}$ attached at points, $T$, $A$, and $R$, which are located on the circle; $m\overset{\frown}{TA} : m\overset{\frown}{AR} : m\overset{\frown}{RT} = 1{:}3{:}5$; $RA = 12$ centimeters; and $PA = 5$ centimeters.

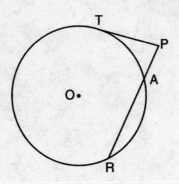

Find the measure of $\angle P$, in degrees, and find the length of rod $\overline{PT}$, to the *nearest tenth of a centimeter*.

34 A surveyor is mapping a triangular plot of land. He measures two of the sides and the angle formed by these two sides and finds that the lengths are 400 yards and 200 yards and the included angle is 50°.

What is the measure of the third side of the plot of land, to the *nearest yard*?

What is the area of this plot of land, to the *nearest square yard*?

# Answers
# June 2004
## Math B

## Answer Key

### PART I

| | | | |
|---|---|---|---|
| **1.** 1 | **6.** 4 | **11.** 4 | **16.** 1 |
| **2.** 1 | **7.** 3 | **12.** 3 | **17.** 4 |
| **3.** 2 | **8.** 1 | **13.** 4 | **18.** 2 |
| **4.** 2 | **9.** 4 | **14.** 3 | **19.** 2 |
| **5.** 3 | **10.** 2 | **15.** 1 | **20.** 3 |

For Parts II, III, and IV, see **Answers Explained** section for computations and methodology to support the given answers.

### PART II

**21.** 41,583

**22.** $\frac{3}{2}$

**23.** $k > \frac{1}{3}$

**24.** $\triangle HOC$ and opposite (with explanation)

**25.** See **Answers Explained** section.

**26.** −17

### PART III

**27.** 44

**28.** 65.27

**29.** $\frac{-3 \pm \sqrt{37}}{7}$

**30.** 3 and 42

**31.** 1,500

**32.** 84

### PART IV

**33.** 80 and 9.2 (See alternative answers in **Answers Explained** section.)

**34.** 312 and 30,642

# Answers Explained

## PART I

**1.** The given expressions are:  $2 - \sqrt{-4} =$  and  $-3 + \sqrt{-16}$

Simplify the square roots in terms of $i = \sqrt{-1}$:  $2 - 2i$ and $-3 + 4i$

Find the sum of the real and imaginary parts:  $-1 + 2i$

The correct choice is (**1**).

**2.** To find the probability that it will rain on *exactly* three of the seven days, use the formula ${}_nC_x p^x (1-p)^{n-x}$, where $n = 7$, $x = 3$, and $p = 0.7$:

$$ {}_nC_x p^x (1-p)^{n-x} = {}_7C_3 (0.7)^3 (1-0.7)^4 = {}_7C_3 (0.7)^3 (0.3)^4 $$

The correct choice is (**1**).

**3.** The given function is $y = \dfrac{2}{3}\sin 4x$. Since the amplitude of any function of the form $y = a \sin bx$ is $a$, the amplitude of the given function is $\dfrac{2}{3}$.

The correct choice is (**2**).

**4.** The graphs of quadratic functions of the form $y = ax^2$ are parabolas that are concave upward when $a > 0$. Note that, since $8 = 2 \cdot 2^2$, the point where $x = 2$ and $y = 8$ satisfies the equation $y = 2x^2$, which is the quadratic function shown in the accompanying graph.

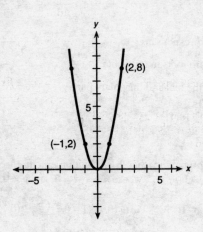

The correct choice is **(2)**.

**5.** For the accompanying graph it is given that the transformation $T$ maps each point $(x,y)$ to point $(2x,4y)$.

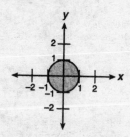

Applying this transformation to points $(0,1)$ and $(1,0)$ from set $A$ yields points $(0,4)$ and $(2,0)$, respectively. Only graph (3) contains these points.

The correct choice is **(3)**.

**6.** The given function is:      $f(x) = 4x^0 + (4x)^{-1}$

Evaluate the function at $x = 4$:      $4(4^0) + (4 \cdot 4)^{-1} = 4(1) + 16^{-1}$

Simplify:      $4 + \dfrac{1}{16} \quad = 4\dfrac{1}{16}$

The correct choice is **(4)**.

**7.** The given function is:      $f(x) = \dfrac{2x^2}{x^2 - 9}$

The domain of this function is the set of all values of $x$ except those that make the denominator equal to 0:      $x^2 - 9 = 0$

$$x^2 = 9$$

$$x = \pm 3$$

The domain of the given function is all real numbers except 3 and –3.

The correct choice is **(3)**.

**8.** An inverse variation is a set of ordered pairs in which the product of the first and second members of each ordered pair is the same nonzero number (stream velocity and distance must have a constant product for each ordered pair).

The graph of an inverse variation is a rectangular hyperbola of two disconnected branches that are asymptotic to the coordinate axes. The first quadrant view of such a graph is shown in graph (1).

The correct choice is **(1)**.

**9.** The given equation is $\log_b x = y$. In a log equation of this form, the base is $b$, the power is $x$, and the exponent is $y$. Therefore the given equation is equivalent to $b^y = x$.

The correct choice is **(4)**.

**10.** The given equation, $i^0 \cdot i^1 \cdot i^2 \cdot i^3 \cdot i^4$, is equal to $i^{10}$. Use the fact that $i^2 = -1$ and $i^4 = 1$; then

$$i^{10} = (i^4)(i^4)(i^2) = (1)(1)(-1) = -1$$

The correct choice is **(2)**.

**11.** The table shows that $y = 5$ when $x = 0$ and doubles each time the value of $x$ increases by 1. Therefore, the $y$ values are generated by the equation $y = 5(2^x)$. Note also that the ordered pair $(1,10)$ from the table satisfies this equation only.

The correct choice is **(4)**.

**12.** It is given that the amount of juice dispensed from a machine is normally distributed with a mean of 10.50 ounces and a standard deviation of 0.75 ounce. This means that 68.2% of the dispensed amounts varies between one standard deviation below the mean and one standard deviation above the mean. Hence 68.2% of the dispensed amounts varies between $10.50 - 0.75 = 9.75$ ounces and $10.50 + 0.75 = 11.25$ ounces, as shown in the accompanying diagram.

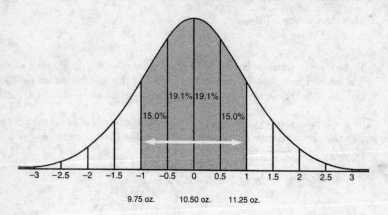

The correct choice is **(3)**.

**13.** It is given that $\theta$ is an acute angle and that $\sin \theta = \dfrac{5}{13}$, as shown in the accompanying diagram. This is a 5-12-13 right triangle, so the leg adjacent to $\theta$ has length 12. Hence, $\cos \theta = \dfrac{12}{13}$.

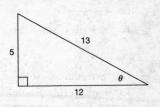

Use the double-angle formula for sine:

$$\sin 2\theta = 2 \sin \theta \cos \theta$$

$$= 2\left(\frac{5}{13}\right)\left(\frac{12}{13}\right)$$

$$= \frac{120}{169}$$

The correct choice is **(4)**.

**14.** Using your graphing calculator, graph each equation. Only $y = -\dfrac{5}{x}$ has a graph that is symmetrical with respect to the origin (180° rotation symmetry), as shown in the accompanying diagram.

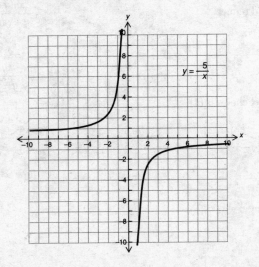

$$y = \frac{5}{x}$$

The correct choice is **(3)**.

**15.** The given complex fraction is:

$$\frac{\dfrac{1}{x}+\dfrac{1}{y}}{\dfrac{1}{x^2}-\dfrac{1}{y^2}}$$

To simplify this fraction, multiply the numerator and denominator by the common denominator of the fraction terms, $x^2y^2$:

$$\frac{x^2y^2\left(\dfrac{1}{x}+\dfrac{1}{y}\right)}{x^2y^2\left(\dfrac{1}{x^2}-\dfrac{1}{y^2}\right)}$$

Distribute, and simplify:

$$\frac{\overset{1}{x^2}y^2\left(\dfrac{1}{x}\right)+x^2\overset{1}{y^2}\left(\dfrac{1}{y}\right)}{x^2y^2\left(\dfrac{1}{x^2}\right)-x^2y^2\left(\dfrac{1}{y^2}\right)}$$

$$\frac{xy^2+x^2y}{y^2-x^2}$$

Factor, and reduce:

$$\frac{xy(y+x)}{(y-x)(y+x)}$$

$$\frac{xy}{y-x}$$

The correct choice is **(1)**.

**16.** It is given that two sides of a triangle are 40 inches and 15 inches and that the angle opposite the 40-inch side measures 120°. Since the angle is obtuse and the longer side is opposite the obtuse angle, only one triangle is possible, as shown in the accompanying diagram.

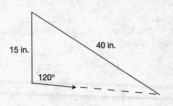

The correct choice is **(1)**.

**17.** The given functions are:

$$f(x) = x + 1 \quad \text{and} \quad g(x) = x^2 - 1$$

To create the composition $(g \circ f)(x)$, substitute $(x^2 + 1)$ for $x$ in the expression for $g(x)$:

$$(g \circ f)(x) = (x + 1)^2 - 1$$

To find $x$ when $(g \circ f)(x)$ equals 0, write the equation:

$$(x + 1)^2 - 1 = 0$$

Simplify:

$$x^2 + 2x + 1 - 1 = 0$$

$$x^2 + 2x = 0$$

Solve the quadratic equation by factoring:

$$x(x + 2) = 0$$

$$x = 0 \quad \text{or} \quad x = -2$$

The correct choice is **(4)**.

**18.** The given equation is: $\sin \theta = a$

Substitute $a$ for $\sin \theta$ in the identity $\sin^2 \theta + \cos^2 \theta = 1$:

$$\sin^2 \theta + \cos^2 \theta = 1$$

$$a^2 + \cos^2 \theta = 1$$

Solve for $\cos \theta$:

$$\cos^2 \theta = 1 - a^2$$

$$\sqrt{\cos^2 \theta} = \sqrt{1 - a^2}$$

$$\cos \theta = \sqrt{1 - a^2}$$

The correct choice is **(2)**.

**19.** The given expression is: $\sqrt[4]{16a^6b^4}$

Express the root using a fractional exponent: $(16a^6b^4)^{\frac{1}{4}}$

Rewrite as an equivalent monomial with each factor raised to the $\frac{1}{4}th$ power: $(16)^{\frac{1}{4}}(a^6)^{\frac{1}{4}}(b^4)^{\frac{1}{4}}$

Simplify: $2a^{\frac{3}{2}}b$

The correct choice is **(2)**.

**20.** It is given that in the accompanying diagram $\overline{HK}$ bisects $\overline{IL}$, making $\overline{IJ} \cong \overline{JL}$, and that $\angle H \cong \angle K$. Since vertical angles $\angle IJH$ and $\angle LJK$ are also congruent, the most direct method to prove $\triangle HIJ \cong \triangle KLJ$ is AAS $\cong$ AAS.

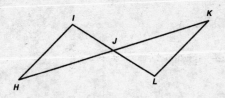

The correct choice is **(3)**.

# PART II

**21.** The given expression for profit during the years 2002, 2003, and 2004, $\sum\limits_{n=0}^{2}(13567n + 294)$, represents the sum of the terms $13567n + 294$ for integer values of $n$ from 0 to 2. To evaluate the summation, substitute each of the values $n = 0$, 1, and 2 into the expression and add the results:

$$\sum_{n=0}^{2}(13567n + 294) = (13567(0) + 294) + (13567(1) + 294) + (13567(2) + 294)$$

$$= 294 \ + 13861 \qquad\qquad + 27428$$

$$= 41583$$

The total profit for the period 2002 to 2004, in dollars, is **41,583**.

**22.** The given equation is an exponential equation since the unknown is in the exponents:

$$27^{2x + 1} = 9^{4x}$$

To solve an exponential equation, rewrite both sides of the equation as powers of the same base:

$$(3^3)^{2x + 1} = (3^2)^{4x}$$
$$3^{3(2x + 1)} = 3^{2(4x)}$$

Set exponents equal to each other, and solve:

$$3(2x + 1) = 2(4x)$$
$$6x + 3 = 8x$$
$$3 = 2x$$
$$\frac{3}{2} = x$$

The solution of the equation $27^{2x + 1} = 9^{4x}$ is $x = \dfrac{3}{2}$.

**23.** The given equation is:

$$3x^2 - 2x + k = 0$$

It is given that the equation has imaginary roots for certain values of $k$. Hence, the value of the discriminant, $b^2 - 4ac$, must be less than zero. Substitute $a = 3$, $b = -2$ and $c = k$:

$$b^2 - 4ac < 0$$

$$(-2)^2 - 4(3)(k) < 0$$

Solve for $k$:

$$4 - 12k < 0$$

$$-12k < 4$$

$$k > \frac{1}{3}$$

The equation $3x^2 - 2x + k = 0$ has imaginary roots for all values of $\boldsymbol{k} > \dfrac{1}{3}$.

**24.** To find the image of $\triangle EOA$ after it is reflected in line $\ell$, reflect each point in line $\ell$. Point $E$ reflects to point $H$, point $O$ reflects to itself, and point $A$ reflects to point $C$, as shown in the accompanying diagram.

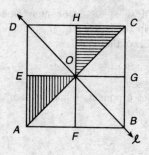

The image of $\triangle EOA$ after it is reflected in line $\ell$ is $\triangle \boldsymbol{HOC}$. The isometry is **opposite** because the orientation of the image, $\triangle HOC$, is not the same as that of the original, $\triangle EOA$.

**25.** It is given that $\overline{AB} \perp \overline{CD}$, as shown in the accompanying diagram.

Plan: Use an indirect proof to show that $\overline{AT}$ *cannot* be perpendicular to $\overline{CD}$ because $\triangle ABT$ cannot have two right angles.

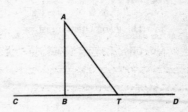

| Statement | Reason |
|---|---|
| 1. $\triangle ABT$, $\overline{CBTD}$, and $\overline{AB} \perp \overline{CD}$ | 1. Given |
| 2. $\angle ABT$ is a right angle. | 2. Perpendicular lines form right angles. |
| 3. Suppose $\overline{AT} \perp \overline{CD}$. | 3. Assumption for indirect proof |
| 4. $\angle ATB$ is a right angle. | 4. Perpendicular lines form right angles. |
| 5. $\triangle ABT$ has two right angles (CONTRADICTION) | 5. A triangle cannot have two right angles. |
| 6. $\overline{AT}$ is *not* perpendicular to $\overline{CD}$. | 6. The contradiction shows that the assumption was false. |

**26.** The given equation is:

$$V = 20\sqrt{C + 273}$$

Substitute $V = 320$, and solve for $C$:

$$320 = 20\sqrt{C + 273}$$

$$\frac{320}{20} = \sqrt{C + 273}$$

$$(16)^2 = \left(\sqrt{C + 273}\right)^2$$

$$256 = C + 273$$

$$-17 = C$$

When the speed of sound is 320 meters per second, the temperature, in degrees Celsius, is **−17**.

## PART III

**27.** The given equation is:

$$L = 6077 - 31 \cos 2\theta$$

Substitute $L = 6076$, and solve for $\theta$:

$$6076 = 6077 - 31 \cos 2\theta$$

$$-1 = -31 \cos 2\theta$$

$$\frac{1}{31} = \cos 2\theta$$

$$\cos^{-1}\left(\frac{1}{31}\right) = 2\theta$$

$$88.1514 \approx 2\theta$$

$$44.0757 \approx \theta$$

To the *nearest degree*, the angle $\theta$, $0° \leq \theta \leq 90°$, at which the length of a nautical mile is approximately 6,076 feet has a measure of **44**.

**28.** It is given that two equal forces act on a body at an angle of 80° and that the resultant force is 100 newtons.

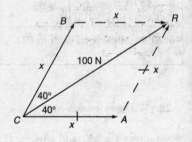

Since the forces are equal, the resultant force forms two isosceles triangles, $\triangle CBR$ and $\triangle CAR$, each with base angles of 40°, as shown in the accompanying diagram. Then the measure of $\angle CAR = 180° - 40° - 40° = 100°$.

To find the value of one of the two equal forces, use the Law of Sines and your calculator:

$$\frac{a}{\sin A} = \frac{b}{\sin B}$$

$$\frac{100}{\sin 100°} = \frac{x}{\sin 40°}$$

$$x = \frac{100 \sin 40°}{\sin 100°}$$

$$\approx 65.2703$$

To the *nearest hundredth of a newton*, the value of one of the two equal forces is **65.27**.

**29.** The given equation is:

$$\frac{4}{x} - \frac{3}{x+1} = 7$$

Multiply each term of the equation by the common denominator, $x(x + 1)$:

$$(x(x+1))\left(\frac{4}{x}\right) - (x(x+1))\left(\frac{3}{x+1}\right) = (x(x+1))(7)$$

$$4(x+1) - 3x = (x^2 + x)7$$

Simplify, and rewrite as a quadratic equation in the form $ax^2 + bx + c = 0$:

$$4x + 4 - 3x = 7x^2 + 7x$$

$$0 = 7x^2 + 6x - 4$$

Solutions of a quadratic equation in the form $ax^2 + bx + c = 0$ are given by the quadratic formula:

$$x = \frac{-b \pm \sqrt{b^2 - 4ac}}{2a}$$

To use the formula to evaluate the roots of the given equation, substitute $a = 7$, $b = 6$, and $c = -4$:

$$= \frac{-6 \pm \sqrt{6^2 - 4 \cdot 7 \cdot -4}}{2 \cdot 7}$$

$$= \frac{-6 \pm \sqrt{36 + 112}}{14}$$

$$= \frac{-6 \pm \sqrt{148}}{14}$$

Simplify the radical:

$$= \frac{-6 \pm \sqrt{4 \cdot 37}}{14}$$

$$= \frac{-6 \pm 2\sqrt{37}}{14}$$

Reduce the fraction by dividing each term by the common factor 2:

$$= \frac{-3 \pm \sqrt{37}}{7}$$

In simplest radical form, the value of $x$ is $\dfrac{-3 \pm \sqrt{37}}{7}$.

**30.** It is given that the ball's height ($y$) at $x$ seconds is modeled by the quadratic equation $y = -16x^2 + 48x + 6$. The graph of a quadratic equation of the form $y = ax^2 + bx + c$ is a parabola. To find when the ball reaches a height of 6 feet, substitute $y = 6$ and solve the equation:

$$y = -16x^2 + 48x + 6$$
$$6 = -16x^2 + 48x + 6$$
$$0 = -16x^2 + 48x$$
$$0 = -16x(x - 3)$$
$$-16x = 0 \quad \text{or} \quad x - 3 = 0$$
$$x = 0 \quad \text{or} \quad x = 3$$

The ball will again be 6 feet above the ground at **3** seconds.

The ball reaches its maximum height at the vertex (turning point) of the parabola. The vertex lies on the axis of symmetry, whose equation is $x = \dfrac{-b}{2a}$. To find the value of $x$ at the turning point, substitute $a = -16$ and $b = 48$:

$$x = \frac{-48}{2(-16)} = \frac{-48}{-32} = \frac{3}{2} \text{ seconds.}$$

To find the maximum height of the ball, substitute $x = \dfrac{3}{2}$ in the equation and solve for $y$:

$$y = -16\left(\frac{3}{2}\right)^2 + 48\left(\frac{3}{2}\right) + 6 = 42$$

The maximum height, in feet, that the ball reaches is **42**.

**31.** It is given that the formula used to determine the age of a specimen is $A = A_0 2^{\frac{-t}{5760}}$, where $A$ is the amount of carbon-14 that the specimen contains, $A_0$ is its original amount of carbon-14 and $t$ is the time, in years. It is also given that the original amount of carbon-14 is 120 milligrams and the current amount is 100 milligrams.

To determine how many years were required for the original amount of carbon-14 to reach the current amount, substitute $A_0 = 120$ and $A = 100$, and solve the equation for $t$:

$$A = A_0 2^{\frac{-t}{5760}}$$

$$100 = 120\left( 2^{\frac{-t}{5760}} \right)$$

$$\log 100 = \log 120 + \left( \frac{-t}{5760} \right) \log 2$$

$$-t = \frac{\log 100 - \log 120}{\log 2} \cdot 5760$$

$$t = -\left( \frac{\log 100 - \log 120}{\log 2} \cdot 5760 \right)$$

Evaluate using your calculator: $\quad t \approx 1515.0782$

To the *nearest hundred years*, the age of this specimen is **1,500**.

**32.** It is given that the sale prices of houses in Mrs. Ramírez's area approximated a normal distribution with a mean of $150,000 and a standard deviation of $25,000. To determine the percentile rank of a sale price of $175,000, first determine how far, in standard deviation units, this price is from the mean:

$$175000 \text{ is } \frac{175000 - 150000}{25000} = 1 \text{ standard deviation above the mean.}$$

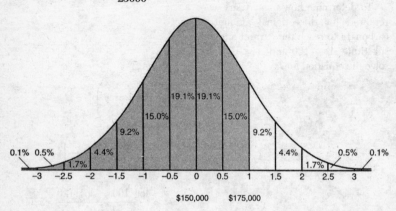

The accompanying figure shows that the percentage of sale prices less than 1 standard deviation above the mean is approximately

$$100 - 9.2 - 4.4 - 1.7 - 0.5 - 0.1 = 84.19$$

To the *nearest whole number*, the percentile rank of a sale price of $175,000 is **84. This percentile means that 84% of the sales prices of the houses sold in this area last month were $175,000 or less.**

It is given that most of the houses sold last month had selling prices between $125,000 and $175,000. As computed earlier, 175,000 is 1 standard deviation above the mean of 150,000. Also,

$$125000 \text{ is } \frac{125000 - 150000}{25000} = -1 = 1 \text{ standard deviation below the mean}$$

**Mrs. Ramírez is correct in saying that most of the houses sold last month had selling prices between $125,0000 and $175,000 because the percentage of sale prices between 1 standard deviation below the mean and 1 standard deviation above the mean is approximately 68.2%.**

# PART IV

**33.**

> NOTE: Because $\overline{PT}$ is not identified as a tangent in this problem, there are alternate solutions that produce different numerical answers each recognized as correct by the Board of Regents. Below is the most common solution, followed by two other solutions also accepted by the Board of Regents.
>
> Consult with your teacher regarding other possible solutions.

**Most Common Answer and Solution**

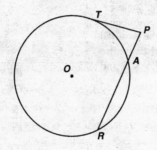

It is given that:

$$\widehat{mTA} : \widehat{mAR} : \widehat{mRT} = 1:3:5$$

Assume that $\widehat{RT}$ is a major arc and therefore forms a complete circle (360°) with $\widehat{AR}$ and $\widehat{TA}$:

$$m\widehat{TA} + m\widehat{AR} + m\widehat{RT} = 360°$$

Substitute $1x$, $3x$, and $5x$ for the arc measures:

$$1x + 3x + 5x = 360$$

Solve for $x$:

$$9x = 360$$

$$x = 40$$

Use the value of $x$ to find the measure of each arc:

$$m\widehat{TA} = x = 40°$$

$$m\widehat{AR} = 3x = 120°$$

$$m\widehat{RT} = 5x = 200°$$

Angle $P$ is formed by secant $\overline{PR}$ and (assumed) tangent $\overline{PT}$, and therefore has a measure equal to half the difference of its intercepted arcs, major arc $\overset{\frown}{RT}$ and arc $\overset{\frown}{TA}$ :

$$m\angle P = \frac{m\overset{\frown}{RT} - m\overset{\frown}{TA}}{2}$$

$$= \frac{200 - 40}{2} = 80$$

The measure, in degrees, of $\angle P$ is **80**.

The Regents allowed the assumption that $\overline{PR}$ is tangent to the circle. When a tangent and a secant are drawn to a circle from the same external point, the tangent segment is the geometric mean of the external segment of the secant and the entire secant. It is given that $PA = 5$ and $RA = 12$. Then:

$$\frac{PA}{PT} = \frac{PT}{PR}$$

$$\frac{5}{PT} = \frac{PT}{17}$$

$$(PT)^2 = 85$$

$$PT \approx 9.2195$$

To the *nearest tenth of a centimeter*, the length of $\overline{PT}$ is **9.2**.

## Alternative Answer and Solution 1 for Finding $PT$

Draw $\triangle TRP$ as shown in the accompanying diagram.

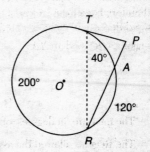

Angle $PTR$ is formed by (assumed) tangent $\overline{PT}$ and chord $\overline{RT}$, and therefore has a measure equal to half its intercepted arc, $\overset{\frown}{TAR}$:

$$m\angle PTR = \frac{m\overset{\frown}{TAR}}{2} = \frac{160}{2} = 80$$

Find $m\angle R$ in $\triangle TRP$:

$$m\angle R = 180 - m\angle P - m\angle PTR$$
$$= 180 - 80 \quad - 80$$
$$= 20$$

It is given that $RA = 12$ and $PA = 5$. Also, $\triangle TPR$ is isosceles ($m\angle P = m\angle PTR$) and therefore $PR = TR$:

$$PR = RA + PA$$
$$= 12 + 5 = 17$$
$$TR = PR$$
$$= 17$$

Use the Law of Cosines to find $PT$:

$$a^2 = b^2 + c^2 - 2ab \cos C$$
$$PT^2 = TR^2 + PR^2 - 2(TR)(PR) \cos R$$
$$= 17^2 + 17^2 - 2(17)(17) \cos 20$$
$$\approx 34.8577$$
$$PT \approx 5.904$$

To the *nearest tenth of a centimeter*, the length of $\overline{PT}$ is **5.9**.

### Alternative Answer and Solution 2 for Finding *PT*

Draw △*TAP* as shown in the accompanying diagram.

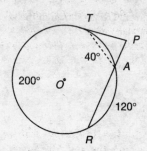

Angle *PTA* is formed by (assumed) tangent $\overline{PT}$ and chord $\overline{AT}$, and therefore has a measure equal to half its intercepted arc, $\stackrel{\frown}{TA}$:

$$m\angle PTA = \frac{m\stackrel{\frown}{TA}}{2}$$

$$= \frac{40}{2} = 20$$

Find m∠*TAP* in △*TAP*:

$$m\angle TAP = 180 - m\angle PTA - m\angle TPA$$

$$= 180 - 20 \qquad - 80$$

$$= 80$$

Use the Law of Sines to find *PT*:

$$\frac{a}{\sin A} = \frac{b}{\sin B}$$

$$\frac{PT}{\sin 80} = \frac{PA}{\sin 20}$$

$$\frac{PT}{\sin 80} = \frac{5}{\sin 20}$$

$$PT = \sin 80 \left( \frac{5}{\sin 20} \right)$$

$$\approx 14.3969$$

To the *nearest tenth of a centimeter*, the length of $\overline{PT}$ is **14.4**.

**34.** It is given that the length of one side of the triangular plot of land is 400 yards, the length of another side is 200 yards, and the included angle is 50°, as shown in the accompanying diagram.

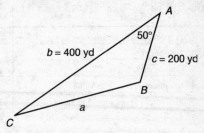

Use the Law of Cosines and your calculator to estimate $a$, the length of the third side of the plot of land:

$$a^2 = b^2 + c^2 - 2bc \cos A$$

$$= 400^2 + 200^2 - 2(400)(200) \cos 50$$

$$\approx 97{,}153.9825$$

$$a \approx 311.70$$

To the *nearest yard*, the measure of the third side of the plot of land is **312**.

To find the area $(K)$ of the plot of land, use the formula:

$$K = \frac{1}{2}bc \sin A$$

Substitute $b = 400$, $c = 200$, and the degree measure (50) of included angle $A$:

$$= \frac{1}{2}(400)(200) \sin 50$$

$$\approx 30{,}641.7778$$

To the *nearest square yard*, the area of the plot is **30,642**.

| Topic | Question Numbers | Number of Points | Your Points |
|---|---|---|---|
| 1. Properties of Real Numbers | — | — | |
| 2. Sequences | — | — | |
| 3. Complex Numbers | 1, 10 | 2 + 2 = 4 | |
| 4. Inequalities, Absolute Value | — | — | |
| 5. Algebraic Expressions, Fractions | 15, 29 | 2 + 4 = 6 | |
| 6. Exponents (zero, negative, rational, scientific notation) | 6, 22 | 2 + 2 = 4 | |
| 7. Radical Expressions | 19, 26 | 2 + 2 = 4 | |
| 8. Quadratic Equations (factors, formula, discriminant) | 23 | 2 | |
| 9. Systems of Equations | — | — | |
| 10. Functions (graphs, domain, range, roots) | 7, 14 | 2 + 2 = 4 | |
| 11. Inverse Functions, Composition | 17 | 2 | |
| 12. Linear Functions | — | — | |
| 13. Parabolas (max/min, axis of symmetry) | 4, 30 | 2 + 4 = 6 | |
| 14. Hyperbolas (including inverse variation) | 8 | 2 | |
| 15. Ellipses | — | — | |
| 16. Exponents and Logarithms | 9, 31 | 2 + 4 = 6 | |
| 17. Trig. (circular functions, unit circle, radians) | 3 | 2 | |
| 18. Trig. Equations and Identities | 18 | 2 | |
| 19. Solving Triangles (sin/cos laws, Pythag. thm., ambig. case) | 13, 16, 27, 28, 34 | 2 + 2 + 4 + 4 + 6 = 18 | |
| 20. Coordinate Geom. (slope, distance, midpoint, circle) | — | — | |
| 21. Transformations, Symmetry | 5, 24 | 2 + 2 = 4 | |
| 22. Circle Geometry | 33 | 6 | |
| 23. Congruence, Similarity, Proportions | — | — | |
| 24. Probability (including Bernoulli events) | 2 | 2 | |
| 25. Normal Curve | 12, 32 | 2 + 4 = 6 | |
| 26. Statistics (center, spread, summation notation) | 21 | 2 | |
| 27. Correlation, Modeling, Prediction | 11 | 2 | |
| 28. Algebraic Proofs, Word Problems | — | — | |
| 29. Geometric Proofs | 20, 25 | 2 + 2 = 4 | |
| 30. Coordinate Geometry Proofs | — | — | |

**Total Raw Score =** _____

## HOW TO CONVERT YOUR RAW SCORE
## TO YOUR MATH B REGENTS EXAMINATION SCORE

Below is the conversion chart that must be used to determine your final score on the June 2004 Regents Examination in Math B. To find your final exam score, locate in the column labeled "Raw Score" the total number of points you scored. Then locate in the adjacent column to the right the scaled score that corresponds to your raw score. The scaled score is your final Math B Regents Examination score.

### Regents Examination in Math B—June 2004
### Chart for Converting Total Test Raw Scores to
### Final Examination Scores (Scaled Scores)

| Raw Score | Scaled Score | Raw Score | Scaled Score | Raw Score | Scaled Score |
|---|---|---|---|---|---|
| 88 | 100 | 58 | 76 | 28 | 46 |
| 87 | 99 | 57 | 76 | 27 | 45 |
| 86 | 98 | 56 | 75 | 26 | 44 |
| 85 | 98 | 55 | 74 | 25 | 42 |
| 84 | 97 | 54 | 73 | 24 | 41 |
| 83 | 96 | 53 | 72 | 23 | 40 |
| 82 | 95 | 52 | 71 | 22 | 38 |
| 81 | 94 | 51 | 70 | 21 | 37 |
| 80 | 93 | 50 | 70 | 20 | 35 |
| 79 | 93 | 49 | 69 | 19 | 34 |
| 78 | 92 | 48 | 68 | 18 | 33 |
| 77 | 91 | 47 | 67 | 17 | 31 |
| 76 | 90 | 46 | 66 | 16 | 29 |
| 75 | 90 | 45 | 65 | 15 | 28 |
| 74 | 89 | 44 | 64 | 14 | 26 |
| 73 | 88 | 43 | 63 | 13 | 25 |
| 72 | 87 | 42 | 62 | 12 | 23 |
| 71 | 87 | 41 | 61 | 11 | 21 |
| 70 | 86 | 40 | 60 | 10 | 19 |
| 69 | 85 | 39 | 59 | 9 | 18 |
| 68 | 84 | 38 | 58 | 8 | 16 |
| 67 | 83 | 37 | 57 | 7 | 14 |
| 66 | 83 | 36 | 56 | 6 | 12 |
| 65 | 82 | 35 | 55 | 5 | 10 |
| 64 | 81 | 34 | 54 | 4 | 8 |
| 63 | 80 | 33 | 52 | 3 | 6 |
| 62 | 80 | 32 | 51 | 2 | 4 |
| 61 | 79 | 31 | 50 | 1 | 2 |
| 60 | 78 | 30 | 49 | 0 | 0 |
| 59 | 77 | 29 | 48 | | |

# Examination August 2004
## Math B

**FORMULAS**

### Area of Triangle

$$K = \frac{1}{2}ab\sin C$$

### Function of the Sum of Two Angles

$$\sin (A + B) = \sin A \cos B + \cos A \sin B$$
$$\cos (A + B) = \cos A \cos B - \sin A \sin B$$

### Function of the Difference of Two Angles

$$\sin (A - B) = \sin A \cos B - \cos A \sin B$$
$$\cos (A - B) = \cos A \cos B + \sin A \sin B$$

### Law of Sines

$$\frac{a}{\sin A} = \frac{b}{\sin B} = \frac{c}{\sin C}$$

### Law of Cosines

$$a^2 = b^2 + c^2 - 2bc \cos A$$

### Functions of the Double Angle

$$\sin 2A = 2 \sin A \cos A$$
$$\cos 2A = \cos^2 A - \sin^2 A$$
$$\cos 2A = 2 \cos^2 A - 1$$
$$\cos 2A = 1 - 2 \sin^2 A$$

## Functions of the Half Angle

$$\sin\frac{1}{2}A = \pm\sqrt{\frac{1-\cos A}{2}}$$

$$\cos\frac{1}{2}A = \pm\sqrt{\frac{1+\cos A}{2}}$$

## Normal Curve
## Standard Deviation

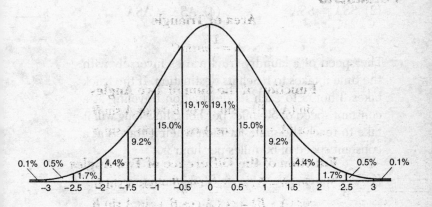

# PART I

Answer all questions in this part. Each correct answer will receive 2 credits. No partial credit will be allowed. For each question, write in the space provided the numeral preceding the word or expression that best completes the statement or answers the question. [40]

1 Which condition does *not* prove that two triangles are congruent?

(1) SSS ≅ SSS      (3) SAS ≅ SAS

(2) SSA ≅ SSA      (4) ASA ≅ ASA     1 _____

2 The speed of a laundry truck varies inversely with the time it takes to reach its destination. If the truck takes 3 hours to each its destination traveling at a constant speed of 50 miles per hour, how long will it take to reach the same location when it travels at a constant speed of 60 miles per hour?

(1) $2\frac{1}{3}$ hours      (3) $2\frac{1}{2}$ hours

(2) 2 hours      (4) $2\frac{2}{3}$ hours     2 _____

3 Which set of ordered pairs is *not* a function?

   (1) {(3,1), (2,1), (1,2), (3,2)}
   (2) {(4,1), (5,1), (6,1), (7,1)}
   (3) {(1,2), (3,4), (4,5), (5,6)}
   (4) {(0,0), (1,1), (2,2), (3,3)}       3 ____

4 A circle has the equation $(x + 1)^2 + (y - 3)^2 = 16$. What are the coordinates of its center and the length of its radius?

   (1) (−1,3) and 4       (3) (−1,3) and 16
   (2) (1,−3) and 4       (4) (1,−3) and 16       4 ____

5 The mean of a normally distributed set of data is 56, and the standard deviation is 5. In which interval do approximately 95.4% of all cases lie?

   (1) 46–56       (3) 51–61
   (2) 46–66       (4) 56–71       5 ____

6 The graph below represents f(x).

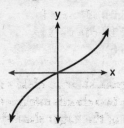

Which graph best represents f(−x)?

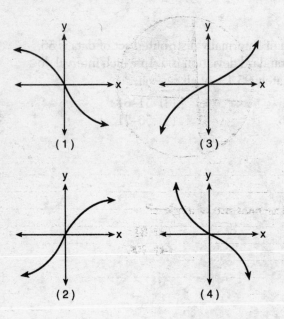

(1)

(3)

(2)

(4)

6 _____

7 When simplified, $i^{27} + i^{34}$ is equal to

   (1) $i$                           (3) $-i - 1$

   (2) $i^{61}$                      (4) $i - 1$            7 _____

8 The accompanying diagram shows a child's spin toy that is constructed from two chords intersecting in a circle. The curved edge of the larger shaded section is one-quarter of the circumference of the circle, and the curved edge of the smaller shaded section is one-fifth of the circumference of the circle.

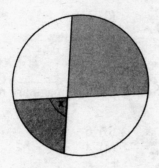

What is the measure of angle $x$?

   (1) $40°$                     (3) $81°$

   (2) $72°$                     (4) $108°$         8 _____

9  If $\sin A = \dfrac{4}{5}$, $\tan B = \dfrac{5}{12}$, and angles $A$ and $B$ are in Quadrant I, what is the value of $\sin (A + B)$?

(1) $\dfrac{63}{65}$          (3) $\dfrac{33}{65}$

(2) $-\dfrac{63}{65}$          (4) $-\dfrac{33}{65}$          9 ____

10  If the tangent of an angle is negative and its secant is positive, in which quadrant does the angle terminate?

(1) I          (3) III
(2) II          (4) IV          10 ____

11  The equation $2x^2 + 8x + n = 0$ has imaginary roots when $n$ is equal to

(1) 10          (3) 6
(2) 8          (4) 4          11 ____

12  What is the middle term in the expansion of $(x + y)^4$?

(1) $x^2y^2$          (3) $6x^2y^2$
(2) $2x^2y^2$          (4) $4x^2y^2$          12 ____

13 What is the image of point (1,1) under $r_{x\text{-axis}} \circ R_{0,90°}$?

   (1) (1,1)               (3) (−1,1)

   (2) (1,−1)            (4) (−1,−1)           13 _____

14 How many distinct triangles can be formed if $m\angle A = 30$, side $b = 12$, and side $a = 8$?

   (1) 1                   (3) 3

   (2) 2                   (4) 0                14 _____

15 The expression $\dfrac{\left(b^{2n+1}\right)^3}{b^n \cdot b^{4n+3}}$ is equivalent to

   (1) $\dfrac{b^n}{2}$              (3) $b^{-3n}$

   (2) $b^n$                (4) $b^{-3n+1}$         15 _____

16 What is the inverse of the function $y = \log_4 x$?

   (1) $x^4 = y$           (3) $4^x = y$

   (2) $y^4 = x$           (4) $4^y = x$          16 _____

17 Which angle is coterminal with an angle of 125°?

   (1) −125°           (3) 235°

   (2) −235°           (4) 425°            17 _____

18 A ball is dropped from a height of 8 feet and allowed to bounce. Each time the ball bounces, it bounces back to half its previous height. The vertical distance the ball travels, $d$, is given by the formula $d = 8 + 16 \sum_{k=1}^{n} \left(\frac{1}{2}\right)^k$, where $n$ is the number of bounces.

Based on this formula, what is the total vertical distance that the ball has traveled after four bounces?

(1) 8.9 ft        (3) 22.0 ft

(2) 15.0 ft       (4) 23.0 ft       18 _____

19 The path traveled by a roller coaster is modeled by the equation $y = 27 \sin 13x + 30$. What is the maximum altitude of the roller coaster?

(1) 13        (3) 30

(2) 27        (4) 57       19 _____

20 The expression $\dfrac{11}{\sqrt{3}-5}$ is equivalent to

(1) $\dfrac{-\sqrt{3}-5}{2}$        (3) $\dfrac{\sqrt{3}-5}{2}$

(2) $\dfrac{-\sqrt{3}+5}{2}$        (4) $\dfrac{\sqrt{3}+5}{2}$       20 _____

## PART II

**Answer all questions in this part. Each correct answer will receive 2 credits. Clearly indicate the necessary steps, including appropriate formula substitutions, diagrams, graphs, charts, etc. For all questions in this part, a correct numerical answer with no work shown will receive only 1 credit.** [12]

21 A ski lift begins at ground level 0.75 mile from the base of a mountain whose face has a 50° angle of elevation, as shown in the accompanying diagram. The ski lift ascends in a straight line at an angle of 20°. Find the length of the ski lift from the beginning of the ski lift to the top of the mountain, to the *nearest hundredth of a mile.*

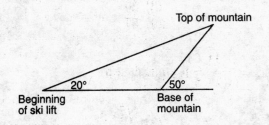

22 Express $\sqrt{-48}$ + 3.5 + $\sqrt{25}$ + $\sqrt{-27}$ in simplest $a + bi$ form.

23 Solve for $x$:    $x^{-3} = \dfrac{27}{64}$

24 The profit a coat manufacturer makes each day is modeled by the equation $P(x) = -x^2 + 120x - 2,000$, where $P$ is the profit and $x$ is the price for each coat sold. For what values of $x$ does the company make a profit? [The use of the accompanying grid is optional.]

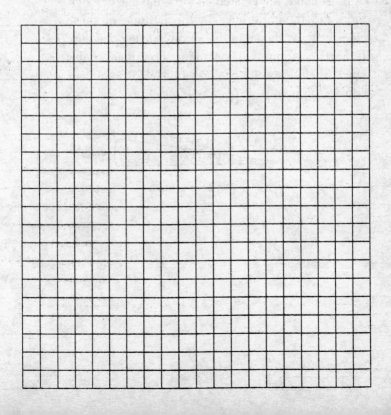

25 Express in simplest form:
$$\dfrac{\dfrac{1}{r} - \dfrac{1}{s}}{\dfrac{r^2}{s^2} - 1}$$

26 Cities $H$ and $K$ are located on the same line of longitude and the difference in the latitude of these cities is 9°, as shown in the accompanying diagram. If Earth's radius is 3,954 miles, how many miles north of city $K$ is city $H$ along arc $HK$? Round your answer to the *nearest tenth of a mile*.

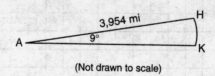

(Not drawn to scale)

## PART III

Answer all questions in this part. Each correct answer will receive 4 credits. Clearly indicate the necessary steps, including appropriate formula substitutions, diagrams, graphs, charts, etc. For all questions in this part, a correct numerical answer with no work shown will receive only 1 credit. [24]

27 A depth finder shows that the water in a certain place is 620 feet deep. The difference between $d$, the actual depth of the water, and the reading is $|d - 620|$ and must be less than or equal to $0.05d$. Find the minimum and maximum values of $d$, to the *nearest tenth of a foot*.

28 An amount of $P$ dollars is deposited in an account paying an annual interest rate $r$ (as a decimal) compounded $n$ times per year. After $t$ years, the amount of money in the account, in dollars, is given by the equation $A = P\left(1 + \dfrac{r}{n}\right)^{nt}$.

Rachel deposited \$1,000 at 2.8% annual interest, compounded monthly. In how many years, to the *nearest tenth of a year*, will she have \$2,500 in the account? [The use of the accompanying grid is optional.]

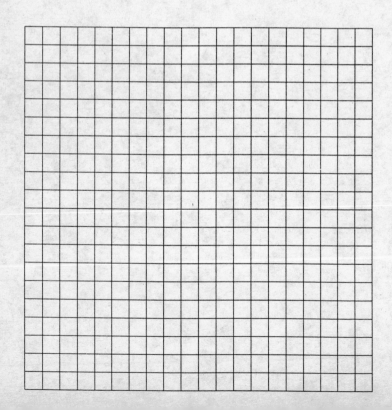

29 A box containing 1,000 coins is shaken, and the coins are emptied onto a table. Only the coins that land heads up are returned to the box, and then the process is repeated. The accompanying table shows the number of trials and the number of coins returned to the box after each trial.

| Trial | 0 | 1 | 3 | 4 | 6 |
|---|---|---|---|---|---|
| Coins Returned | 1,000 | 610 | 220 | 132 | 45 |

Write an exponential regression equation, rounding the calculated values to the *nearest ten-thousandth*.

Use the equation to predict how many coins would be returned to the box after the eighth trial.

30  Tim Parker, a star baseball player, hits one home run for every ten times he is at bat. If Parker goes to bat five times during tonight's game, what is the probability that he will hit *at least* four home runs?

31  A rectangular piece of cardboard is to be formed into an uncovered box. The piece of cardboard is 2 centimeters longer than it is wide. A square that measures 3 centimeters on a side is cut from each corner. When the sides are turned up to form the box, its volume is 765 cubic centimeters. Find the dimensions, in centimeters, of the original piece of cardboard.

32  Solve algebraically for all values of $\theta$ in the interval $0° \leq \theta < 360°$ that satisfy the equation $\dfrac{\sin^2 \theta}{1 + \cos \theta} = 1.$

## PART IV

Answer all questions in this part. Each correct answer will receive 6 credits. Clearly indicate the necessary steps, including appropriate formula substitutions, diagrams, graphs, charts, etc. For all questions in this part, a correct numerical answer with no work shown will receive only 1 credit.  [12]

33 The tide at a boat dock can be modeled by the equation $y = -2 \cos\left(\frac{\pi}{6} t\right) + 8$, where $t$ is the number of hours past noon and $y$ is the height of the tide, in feet. For how many hours between $t = 0$ and $t = 12$ is the tide at least 7 feet? [The use of the accompanying grid is optional.]

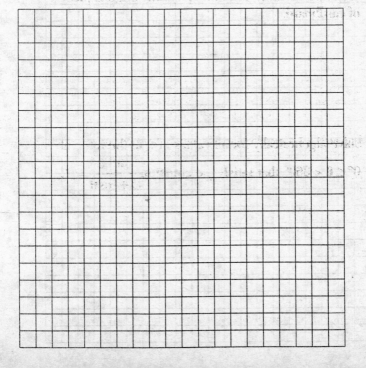

34 The coordinates of quadrilateral *JKLM* are *J*(1,–2),
*K*(13,4), *L*(6,8), and *M*(–2,4). Prove that quadrilateral *JKLM* is a trapezoid but *not* an isosceles
trapezoid. [The use of the accompanying grid is
optional.]

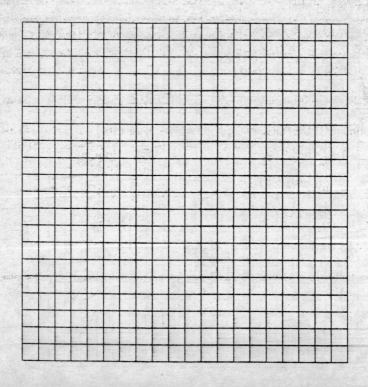

# Answers August 2004

## Math B

## Answer Key

### PART I

| | | | |
|---|---|---|---|
| **1.** 2 | **6.** 4 | **11.** 1 | **16.** 3 |
| **2.** 3 | **7.** 3 | **12.** 3 | **17.** 2 |
| **3.** 1 | **8.** 3 | **13.** 4 | **18.** 4 |
| **4.** 1 | **9.** 1 | **14.** 2 | **19.** 4 |
| **5.** 2 | **10.** 4 | **15.** 2 | **20.** 1 |

For Parts II, III, and IV, see Answers Explained section for computations and methodology to support the given answers.

### PART II

**21.** 1.15

**22.** $8.5 + 7i\sqrt{3}$

**23.** $\frac{4}{3}$ or $1\frac{1}{3}$ or $1.\overline{3}$

**24.** $20 < x < 100$

**25.** $-\dfrac{s}{r(r+s)}$ or $-\dfrac{s}{r^2+rs}$

**26.** 621.1

### PART III

**27.** 590.5 and 652.6

**28.** 32.8

**29.** $y = 1,018.2839(0.5969)^x$ and 16

**30.** .00046 or $\dfrac{46}{100,000}$

**31.** 21 by 23

**32.** 90 and 270

### PART IV

**33.** 8

**34.** $\overline{JK} \parallel \overline{ML}$, $\overline{MJ} \not\parallel \overline{KL}$, and $\overline{MJ} \neq \overline{KL}$

# Answers Explained

## PART I

1. If two sides and the included angle of one triangle are congruent to two sides and the included angle of another triangle (SAS $\cong$ SAS), then the triangles must be congruent. The SSS $\cong$ SSS and ASA $\cong$ ASA conditions also prove that two triangles are congruent. However, if two sides and a nonincluded angle of one triangle are congruent to two sides and a nonincluded angle of another triangle (SSA $\cong$ SSA), the triangles are *not* necessarily congruent, as demonstrated in the accompanying diagram.

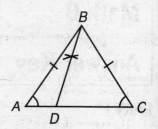

$$\overline{BD} \cong \overline{BD} \qquad \text{(S)}$$
$$\overline{AB} \cong \overline{BC} \qquad \text{(S)}$$
$$\angle A \cong \angle C \qquad \text{(A)}$$
$$\triangle ABD \not\cong \triangle CBD$$

The correct choice is **(2)**.

2. It is given that the speed of a laundry truck varies inversely with the time it takes to reach its destination. Let $S$ = the speed and $T$ = the length of time.

If two variables vary inversely, then their product is a constant. Call this constant $k$:

$$ST = k$$

It is given that the time required for the truck to reach its destination is 3 hours when its speed is 50 miles per hour. Substitute $S = 50$ and $T = 3$ to find the value of $k$:

$$(50)(3) = k$$
$$150 = k$$

To find the time required when the truck's speed is 60 miles per hour, substitute $S = 60$ and find the value of $T$:

$$60T = 150$$
$$T = 2.5$$

The time required is $2\frac{1}{2}$ hours.

The correct choice is **(3)**.

**3.** A function is a relation in which no two ordered pairs have the same first member and different second members. Choice (1) contains the ordered pairs (3,1) and (3,2) and therefore is *not* a function.

The correct choice is **(1)**.

**4.** The given equation of a circle is $(x + 1)^2 + (y - 3)^2 = 16$. The general equation of a circle with a radius of length $r$ and center at $(h,k)$ is $(x - h)^2 + (y - k)^2 = r^2$.

In the given equation, $r^2 = 16$ so $r = 4$. Also, $h = -1$ and $k = 3$. Therefore the coordinates of the center of the given circle are $(-1,3)$ and the length of its radius is 4 units.

The correct choice is **(1)**.

**5.** It is given that in a normally distributed set of data the mean is 56 and the standard deviation is 5. In a normal distribution, 95.4% of the cases lie between two standard deviations below the mean and two standard deviations above the mean. Therefore, 94.5% of the cases in this distribution lie between $56 - 2(5) = 46$ and $56 + 2(5) = 66$.

The correct choice is **(2)**.

**6.** The graph of $f(-x)$ is the reflection of the graph of $f(x)$ in the $y$-axis, as shown in the accompanying diagram. This graph is shown in choice (4).

The correct choice is **(4)**.

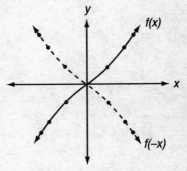

**7.** The given expression is: $i^{27} + i^{34}$

Since $i^2 = -1$, express the given expression in terms of $i^2$: $(i^2)^{13}(i) + (i^2)^{17}$

Substitute $-1$ for $i^2$: $(-1)^{13}(i) + (-1)^{17}$

Simplify: $(-1)(i) + (-1)$

$-i - 1$

The correct choice is **(3)**.

**8.** It is given that the curved edge of the larger shaded section is one-quarter of the circumference of the circle and that the curved edge of the smaller shaded section is one-fifth of the circumference of the circle. Therefore, the measure of the arc of the larger shaded section is $\left(\dfrac{1}{4}\right)(360) = 90°$, and the measure of the arc of the smaller shaded section is $\left(\dfrac{1}{5}\right)(360) = 72°$.

Angle $x$ is formed by two intersecting chords and therefore has a measure equal to one-half the sum of the measures of its intercepted arcs, $90°$ and $72°$. The measure of angle $x = \dfrac{90° + 72°}{2} = 81°$.

The correct choice is **(3)**.

**9.** It is given that angles $A$ and $B$ are in Quadrant I; therefore they are acute angles. It is also given that $\sin A = \dfrac{4}{5}$ and $\tan B = \dfrac{5}{12}$, as shown in the accompanying diagram.

The triangle containing angle $A$ is a 3-4-5 right triangle, so the leg adjacent to $A$ has a length of 3. Hence $\cos A = \dfrac{3}{5}$.

The triangle containing angle $B$ is a 5-12-13 right triangle, so the leg adjacent to $B$ has a length of 12. Hence $\sin B = \dfrac{5}{13}$ and $\cos B = \dfrac{12}{13}$.

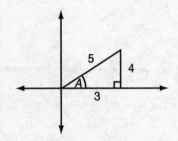 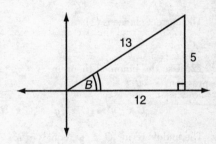

Use the formula for the sine of the sum of two angles:

$$\sin (A + B) = \sin A \cos B + \cos A \sin B$$
$$= \left(\frac{4}{5}\right)\left(\frac{12}{13}\right) + \left(\frac{3}{5}\right)\left(\frac{5}{13}\right)$$
$$= \left(\frac{48}{65}\right) + \left(\frac{15}{65}\right)$$
$$= \frac{63}{65}$$

The correct choice is **(1)**.

**10.** It is given that the tangent of an angle is negative and its secant is positive. Tangent is negative for angles that terminate in Quadrant II or IV. Secant is the reciprocal of cosine and is positive for angles that terminate in Quadrant I or IV. Therefore an angle with a negative tangent and a positive secant must terminate in Quadrant IV.

The correct choice is **(4)**.

**11.** The given equation is: $\qquad\qquad 2x^2 + 8x + n = 0$

A quadratic equation in the form $ax^2 + bx + c = 0$ has imaginary roots when the value of the discriminant, $b^2 - 4ac$, is less than 0. Substitute $a = 2$, $b = 8$, and $c = n$:

$$b^2 - 4ac < 0$$
$$(8)^2 - 4(2)(n) < 0$$
$$64 - 8n < 0$$
$$64 < 8n$$
$$n > 8$$

Only one choice gives a value for $n$ greater than 8.

The correct choice is **(1)**.

**12.** Use the binomial theorem, $(x + y)^n = \sum_{r=0}^{n} {}_nC_r x^{n-r}y^r$, to expand the given expression, $(x + y)^4$:

$${}_4C_0x^{4-0}y^0 + {}_4C_1x^{4-1}y^1 + {}_4C_2x^{4-2}y^2 + {}_4C_3x^{4-3}y^3 + {}_4C_4x^{4-4}y^4$$

The middle term, ${}_4C_2x^{4-2}y^2$, equals $6x^2y^2$.

The correct choice is **(3)**.

**13.** The given point is: $\qquad\qquad (1,1)$

Since $R_{0,90°}(x,y) = (-y,x)$: $\qquad\qquad R_{0,90°}(1,1) = (-1,1)$

Since $r_{x\text{-}axis}(x,y) = (x,-y)$: $\qquad\qquad r_{x\text{-}axis}(-1,1) = (-1,-1)$

The correct choice is **(4)**.

**14.** It is given that $m\angle A = 30$, side $b = 12$, and side $a = 8$, as shown in the accompanying diagram.

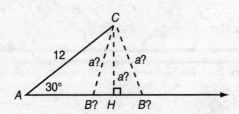

To determine how many distinct triangles can be formed, find the shortest distance from $C$ to the opposite side of the triangle, which is the length of the perpendicular segment $\overline{CH}$. Use the sine function and your calculator:

$$\sin 30 = \frac{CH}{12}$$
$$CH = 12 \sin 30$$
$$= 6$$

Since $a = 8$ is longer than $\overline{CH}$ and shorter than $\overline{AC}$, point $B$ can be at either of two different locations along $\overrightarrow{AH}$ : between $A$ and $H$ or beyond $H$, as shown in the diagram. Hence two distinct triangles can be formed.

The correct choice is **(2)**.

**15.** The given expression is:

$$\frac{\left(b^{2n+1}\right)^{3}}{b^{n} \cdot b^{4n+3}}$$

Apply the rule $(a^n)^m = a^{nm}$ to the numerator and the rule $a^n \cdot a^m = a^{n+m}$ to the denominator:

$$\frac{b^{6n+3}}{b^{5n+3}}$$

Apply the rule $\dfrac{a^n}{a^m} = a^{n-m}$, and simplify:

$$b^{(6n+3)-(5n+3)} = b^{6n+3-5n-3} = b^{n}$$

The correct choice is **(2)**.

**16.** The given function is: $\qquad y = \log_4 x$

To find the inverse of a function, interchange
$x$ and $y$: $\qquad\qquad\qquad\qquad\qquad x = \log_4 y$

In this log equation, the base is 4, the exponent
is $x$, and the power is $y$: $\qquad\qquad\qquad 4^x = y$

The correct choice is **(3)**.

**17.** The terminal side of an angle of 125° lies in Quadrant II and forms a reference angle of 180° − 125° = 55° with the $x$-axis. Of the choices given, only an angle of −235° terminates in Quadrant II and has a reference angle of 55°.

The correct choice is **(2)**.

**18.** The given equation for the vertical distance the ball travels after $n$ bounces, $d = 8 + 16 \sum\limits_{k=1}^{n} \left(\dfrac{1}{2}\right)^k$, represents 8 plus 16 times the sum of terms $\left(\dfrac{1}{2}\right)^k$ for integer values of $k$ from 1 to $n$. To find the total vertical distance the ball has traveled after four bounces, substitute each of the values $k = 1, 2, 3$, and 4 into the expression and add the results:

$$8 + 16 \sum_{k=1}^{4} \left(\frac{1}{2}\right)^k = 8 + 16\left(\left(\frac{1}{2}\right)^1 + \left(\frac{1}{2}\right)^2 + \left(\frac{1}{2}\right)^3 + \left(\frac{1}{2}\right)^4\right)$$

$$= 8 + 16\left(\frac{1}{2} + \frac{1}{4} + \frac{1}{8} + \frac{1}{16}\right)$$

$$= 8 + 16\left(\frac{8}{16} + \frac{4}{16} + \frac{2}{16} + \frac{1}{16}\right) = 8 + 16\left(\frac{15}{16}\right) = 8 + 15 = 23.0 \text{ ft}$$

The correct choice is **(4)**.

**19.** It is given that the path traveled by a roller coaster is modeled by the equation $y = 27 \sin 13x + 30$. The graph of an equation in the form $y = a \sin bx + c$ has an amplitude of $a$ and is translated vertically $c$ units from the $x$-axis. (The maximum altitude modeled by this equation can also be seen on a graph of this equation produced by a graphing calculator.) Therefore the maximum altitude of the roller coaster is $27 + 30 = 57$ units.

The correct choice is **(4)**.

**20.** The given expression is:

$$\frac{11}{\sqrt{3} - 5}$$

Multiply the numerator and the denominator by the conjugate of the denominator, $\sqrt{3} + 5$:

$$\left(\frac{11}{\sqrt{3} - 5}\right) \cdot \left(\frac{\sqrt{3} + 5}{\sqrt{3} + 5}\right)$$

Simplify:

$$\frac{11\left(\sqrt{3} + 5\right)}{3 - 25}$$

$$\frac{11\left(\sqrt{3} + 5\right)}{-22}$$

Divide the numerator and the denominator by a common factor of $-11$:

$$\frac{\overset{-1}{\cancel{11}}\left(\sqrt{3} + 5\right)}{\underset{2}{\cancel{-22}}}$$

Simplify:

$$\frac{-\sqrt{3} - 5}{2}$$

The correct choice is **(1)**.

# PART II

**21.** The given information indicates that the ski lift forms a triangle with the side of a mountain, as shown in the accompanying diagram. It is also given that the distance between the beginning of the lift and the base of the mountain is 0.75 mile and that the lift ascends in a straight line at an angle of 20°.

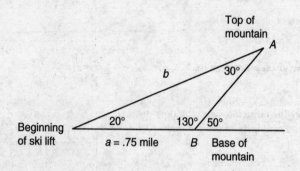

The angle opposite the length of the ski lift has a measure of 180° − 50° = 130°, and the angle opposite the ground has a measure of 180° − 130° − 20° = 30°. To find the length of the lift, use the Law of Sines and your calculator:

$$\frac{a}{\sin A} = \frac{b}{\sin B}$$

$$\frac{0.75}{\sin 30°} = \frac{b}{\sin 130°}$$

$$b = \frac{0.75(\sin 130°)}{\sin 30°}$$

$$\approx 1.1491 \text{ miles}$$

To the *nearest hundredth of a mile*, the length of the ski lift from the beginning of the ski lift to the top of the mountain is **1.15**.

**22.** The given expression is:  $\sqrt{-48} + 3.5 + \sqrt{25} + \sqrt{-27}$

Express negative radicals in terms of $i$:  $i\sqrt{48} + 3.5 + \sqrt{25} + i\sqrt{27}$

Simplify the radicals:  $i\sqrt{16 \cdot 3} + 3.5 + 5 + i\sqrt{9 \cdot 3}$

$$4i\sqrt{3} + 3.5 + 5 + 3i\sqrt{3}$$

Combine like terms:  $8.5 + 7i\sqrt{3}$

The given expression in simplest $a + bi$ form is **$8.5 + 7i\sqrt{3}$**.

**23.** The given equation is:  $x^{-3} = \dfrac{27}{64}$

Raise both sides of the equation to the $-1$ power:  $(x^{-3})^{-1} = \left(\dfrac{27}{64}\right)^{-1}$

Simplify:  $x^3 = \dfrac{64}{27}$

Take the cube root of each side:  $\sqrt[3]{x^3} = \sqrt[3]{\dfrac{64}{27}}$

$$x = \dfrac{4}{3}$$

The value of $x$ is $\dfrac{4}{3}$ or $1\dfrac{1}{3}$ or $1.\overline{3}$.

**24.** It is given that daily profit is modeled by the equation $P(x) = -x^2 + 120x - 2000$, where $P$ is the profit and $x$ is the price for each coat sold.

To find the values of $x$ for which the company makes a profit, solve the inequality:

$$-x^2 + 120x - 2000 > 0$$

Multiply both sides by $-1$:

$$x^2 - 120x + 2000 < 0$$

Factor:

$$(x - 20)(x - 100) < 0$$

If the product of two terms is negative, then exactly one of the terms is negative:

$(x - 20) < 0$ and $(x - 100) > 0$     or     $(x - 20) > 0$ and $(x - 100) < 0$

$x < 20$ and $x > 100$     or     $x > 20$ and $x < 100$

REJECT: There are no numbers less than 20 and greater than 100.

$$20 < x < 100$$

The values of $x$ for which the company makes a profit are **$20 < x < 100$**.

Graphical Solution:

Plot the equation $y = -x^2 + 120x - 2000$ on a graphing calculator and draw an accurate graph on the grid by carefully plotting points using the coordinates obtained from the trace or table feature of the calculator. The equation, axes, scales, and the points where the graph intercepts the $x$ axis, (20,0) and (100,0), must be labeled.

| Price per Coat ($) | Profit ($) |
|---|---|
| 0 | −2000 |
| 10 | −900 |
| 20 | 0 |
| 30 | 700 |
| 40 | 1200 |
| 50 | 1500 |
| 60 | 1600 |
| 70 | 1500 |
| 80 | 1200 |
| 90 | 700 |
| 100 | 0 |
| 110 | −900 |
| 120 | −2000 |

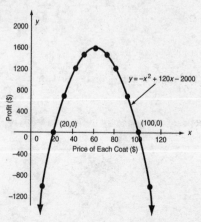

The profit equation is above 0 for values of $x$ greater than 20 and less than 100. The values of $x$ for which the company makes a profit are **$20 < x < 100$**.

**25.** The given expression is:

$$\dfrac{\dfrac{1}{r} - \dfrac{1}{s}}{\dfrac{r^2}{s^2} - 1}$$

Multiply each term by the common denominator, $rs^2$:

$$\dfrac{(rs^2)\dfrac{1}{r} - (rs^2)\dfrac{1}{s}}{(rs^2)\dfrac{r^2}{s^2} - (rs^2)1}$$

Cancel common factors:

$$\dfrac{(rs^2)\dfrac{1}{\cancel{r}} - (rs^{\cancel{2}})\dfrac{1}{\cancel{s}}}{(r\cancel{s^2})\dfrac{r^2}{\cancel{s^2}} - (rs^2)1}$$

Simplify:

$$\dfrac{s^2 - rs}{r^3 - rs^2}$$

Factor:

$$\dfrac{s(s-r)}{r(r^2 - s^2)}$$

$$\dfrac{s(s-r)}{r(r-s)(r+s)}$$

Apply the fact that
$$\dfrac{s-r}{r-s} = \dfrac{-1(\cancel{-s+r})}{\cancel{r-s}} = -1:$$

$$\dfrac{s(\overset{-1}{\cancel{s-r}})}{r(\cancel{r-s})(r+s)} = -\dfrac{s}{r(r+s)} \text{ or } -\dfrac{s}{r^2 + rs}$$

The given expression in simplest form is $-\dfrac{\boldsymbol{s}}{\boldsymbol{r(r+s)}}$ or $-\dfrac{\boldsymbol{s}}{\boldsymbol{r^2 + rs}}$.

**26.** The given information indicates that cities $H$ and $K$ lie on the circumference of circle $A$ formed by radius $\overline{AH}$, as shown in the accompanying diagram. Points $H$, $A$, and $K$ form a sector of the circle with a central angle of 9°; therefore arc $HK$ has a length equal to $\dfrac{9}{360}$ of the circumference of circle $A$.

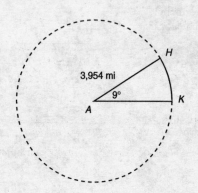

The circumference of circle $A = 2\pi r = 2\pi(3954) = 7908\pi$. Therefore, the length of arc $HK = \left(\dfrac{9}{360}\right)(7908\pi) \approx 621.0929$.

To the *nearest tenth of a mile*, city $H$ is **621.1** miles north of city $K$ along arc $HK$.

# PART III

**27.** It is given that $|d - 620|$ must be less than or equal to $0.05d$:

$$|d - 620| \leq 0.05d$$

Rewrite the inequality without the absolute-value expression:

$$d - 620 \leq 0.05d \quad \text{or} \quad d - 620 \geq -0.05d$$

Solve each inequality:

$$0.95d \leq 620 \quad \text{or} \quad 1.05d \geq 620$$

$$d \leq \frac{620}{0.95} \quad \text{or} \quad d \geq \frac{620}{1.05}$$

The solution set is $\{d \mid 590.4762 \leq d \leq 652.6316\}$.

To the *nearest tenth of a foot*, the minimum and maximum values of $d$ are **590.5** and **652.6**.

**28.** It is given that, after $t$ years, the amount of money in an account paying annual interest is given by the equation $A = P\left(1+\dfrac{r}{n}\right)^{nt}$ , where $A$ is the amount of money, $P$ is the amount deposited, $r$ is the interest rate (as a decimal), and $n$ is the number of times per year the interest is compounded.

To determine how many years it will take the $1,000 that Rachel invested at 2.8% annual interest, compounded monthly, to reach $2,500, substitute $A = 2500$, $P = 1000$, $r = 0.028$, and $n = 12$, and solve the equation for $t$:

$$A = P\left(1+\frac{r}{n}\right)^{nt}$$

$$2500 = 1000\left(1+\frac{0.028}{12}\right)^{12t}$$

$$\log 2500 = \log\left[1000\left(1+\frac{0.028}{12}\right)^{12t}\right]$$

$$\log 2500 = \log 1000 + 12t\left[\log\left(1+\frac{0.028}{12}\right)\right]$$

$$\frac{\log 2500 - \log 1000}{\log\left(1+\dfrac{0.028}{12}\right)} = 12t$$

$$393.1539954 \approx 12t$$

$$32.7628 \approx t$$

To the *nearest tenth of a year*, Rachel's account will total $2,500 in **32.8** years.

**29.** To find the exponential regression equation required to model the given (trial number, coins returned) data, use a calculator. Enter each trial number as $x$ and the corresponding coins-returned value as $y$, and perform an exponential regression.

The equation, with the regression coefficients rounded to the *nearest ten-thousandth*, is

$$y = 1018.2839(0.5969)^x$$

To predict how many coins would be returned to the box after the eighth trial, substitute $x = 8$ in the regression equation:

$$y = 1018.2839(0.5969)^8 \approx 16.4090$$

The number of coins that the equation predicts would be returned to the box after the eighth trial is **16**.

**30.** To find the probability that Parker will hit *at least* four home runs in the five times he goes to bat, use the formula $_nC_x p^x(1 - p)^{n-x}$. Compute first the probability that Parker will hit four out of five home runs ($n = 5$, $p = \dfrac{1}{10}$, and $x = 4$), and then the probability that he will hit five out of five home runs ($n = 5$, $p = \dfrac{1}{10}$, and $x = 5$). Finally, add these probabilities together:

$$_5C_4 0.1^4(1 - 0.1)^{5-4} + {}_5C_5 0.1^5(1 - 0.1)^{5-5} = .00046$$

The probability that Parker will hit at least four home runs in the five times he goes to bat is **.00046** or $\dfrac{46}{100,000}$.

**31.** It is given that a rectangular piece of cardboard is 2 centimeters longer than it is wide and that a square 3 centimeters on a side is cut from each corner. The resulting piece of cardboard has side lengths of $x - 4$, 3, and $x - 6$, as shown in the accompanying diagram.

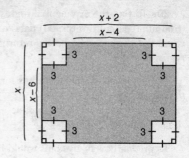

The uncovered box formed by the cardboard, with dimensions of height 3, length $x - 4$, and width $x - 6$, is shown in the accompanying diagram.

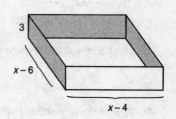

The formula for the volume of a rectangular prism is:

$$V = LWH$$

It is given that the volume of the box is 765 cubic centimeters. Substitute $V = 765$, $L = x - 4$, $W = x - 6$, and $H = 3$:

$$765 = (x - 4)(x - 6)(3)$$

Simplify:

$$765 = (x^2 - 10x + 24)(3)$$

$$255 = x^2 - 10x + 24$$

To solve the quadratic equation, rewrite it in standard form with 0 on one side and all other terms on the other side:

$$0 = x^2 - 10x - 231$$

Factor the quadratic trinomial as the product of two binomials:

$$0 = (x - 21)(x + 11)$$

If the product of two quantities is 0, then at least one of them is 0. Solve the two equations for $x$:

$$0 = x - 21 \quad \text{or} \quad 0 = x + 11$$
$$x = 21 \quad\quad\quad \text{or} \quad\quad x = -11$$

Substitute $x = 21$ and $x = -11$ into $x$, $x - 4$, and $x - 6$, the expressions for the dimensions of the box:

For $x = 21$:    $L = x - 4 = 21 - 4 = 17$
                  $W = x - 6 = 21 - 6 = 15$

For $x = -11$:   $L = x - 4 = -11 - 4 = -15$
                  $W = x - 6 = -11 - 6 = -17$

The dimensions of a box cannot be negative, so reject $x = -11$.

Substitute $x = 21$ into the expressions $x$ and $x + 2$, the dimensions of the **original** rectangle:

$x = 21$
$x + 2 = 21 + 2 = 23$

The dimensions, in centimeters, of the original piece of cardboard are **21 by 23**.

**32.** The given equation is:

$$\frac{\sin^2 \theta}{1 + \cos \theta} = 1$$

Multiply both sides of the equation by the common denominator, $1 + \cos \theta$, and simplify:

$$(1 + \cos \theta) \frac{\sin^2 \theta}{1 + \cos \theta} = 1(1 + \cos \theta)$$

$$\sin^2 \theta = 1 + \cos \theta$$

Substitute $\sin^2 \theta + \cos^2 \theta$ for 1:

$$\sin^2 \theta = \sin^2 \theta + \cos^2 \theta + \cos \theta$$

Subtract $\sin^2 \theta$ from both sides of the equation:

$$0 = \cos^2 \theta + \cos \theta$$

Factor the equation:

$$0 = \cos \theta (\cos \theta + 1)$$

If the product of two quantities is 0, then at least one of them is 0. Solve the two equations for $\theta$, $0° \le \theta < 360°$:

$0 = \cos \theta$     or     $0 = \cos \theta + 1$
                         $-1 = \cos \theta$
$\theta = 90°, 270°$   or   $\theta = 180°$

Check each value of $\theta$ by substituting into the given equation. The equation $\theta = 180°$ makes the denominator of the given equation, $1 + \cos \theta$, equal to zero and therefore must be rejected.

The values of $\theta$ in the interval $0° \le \theta < 360°$ that satisfy the equation are **90** and **270**.

## PART IV

**33.** It is given the tide at a boat dock can be modeled by the equation $y = -2\cos\left(\dfrac{\pi}{6}t\right) + 8$, where $t$ is the number of hours past noon and $y$ is the height of the tide, in feet. To find how many hours between $t = 0$ and $t = 12$ the tide is at least 7 feet, solve the inequality $-2\cos\left(\dfrac{\pi}{6}t\right) + 8 \geq 7$:

$$-2\cos\left(\frac{\pi}{6}t\right) + 8 \geq 7$$

$$-2\cos\left(\frac{\pi}{6}t\right) \geq -1$$

$$\cos\left(\frac{\pi}{6}t\right) \leq \frac{1}{2}$$

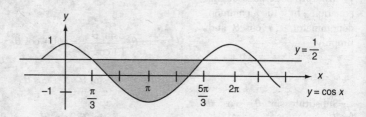

As shown in the accompanying graph of $y = \cos x$ and $y = \dfrac{1}{2}$, cosine is less than or equal to $\dfrac{1}{2}$ for angles between $\dfrac{\pi}{3}$ and $\dfrac{5\pi}{3}$ radians. Therefore, solve the inequality $\dfrac{\pi}{3} \leq \dfrac{\pi}{6}t \leq \dfrac{5\pi}{3}$:

$$\frac{\pi}{3} \leq \frac{\pi}{6}t \leq \frac{5\pi}{3}$$

$$\overset{2}{\cancel{6}}\left(\frac{\pi}{\cancel{3}}\right) \leq \cancel{6}\left(\frac{\pi}{\cancel{6}}t\right) \leq \overset{2}{\cancel{6}}\left(\frac{5\pi}{\cancel{3}}\right)$$

$$2\pi \leq \pi t \leq 10\pi$$

$$2 \leq t \leq 10$$

The height of the tide is at least 7 feet when $t$ is between 2 and 10, and $10 - 2 = 8$.

The number of hours between $t = 0$ and $t = 12$ that the tide is at least 7 feet is **8**.

Graphical Solution:

Plot the graph of $y = -2 \cos\left(\dfrac{\pi}{6}t\right) + 8$ in *radian mode* on a graphing calculator.

Draw an accurate graph on the grid by carefully plotting points using the coordinates obtained from the trace or table feature of the calculator. Also plot the graph of the line $y = 7$ to mark when the tide is 7 feet high. The equations, axes, scales, and the points where the graphs intersect, $(2,7)$ and $(10,7)$, must be labeled.

| Hours Past Noon | Height of Tide (ft) |
|---|---|
| 0 | 6.00000 |
| 1 | 6.26795 |
| 2 | 7.00000 |
| 3 | 8.00000 |
| 4 | 9.00000 |
| 5 | 9.73205 |
| 6 | 10.00000 |
| 7 | 9.73205 |
| 8 | 9.00000 |
| 9 | 8.00000 |
| 10 | 7.00000 |
| 11 | 6.26795 |
| 12 | 6.00000 |

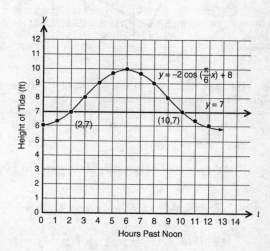

The height of the tide is at least 7 feet when t is between 2 and 10, and $10 - 2 = 8$.

The number of hours between $t = 0$ and $t = 12$ that the tide is at least 7 feet is **8**.

**34.** It is given that the coordinates of the vertices of quadrilateral *JKLM* are $J(1,-2)$, $K(13,4)$, $L(6,8)$ and $M(-2,4)$ as shown in the accompanying diagram.

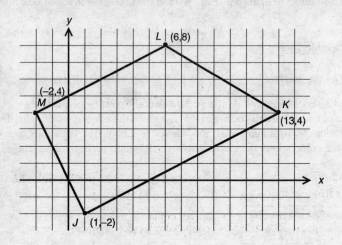

The slope of a segment whose endpoints are $(x_1, y_1)$ and $(x_2, y_2)$ can be determined by using the slope formula:

$$\text{slope} = \frac{y_2 - y_1}{x_2 - x_1}.$$

Find the slopes of all four sides of the quadrilateral:

- To find the slope of $\overline{JK}$, let $(x_1, y_1) = J(1,-2)$ and $(x_2, y_2) = K(13,4)$:

$$\text{Slope of } \overline{JK} = \frac{4 - (-2)}{13 - 1} = \frac{6}{12} = \frac{1}{2}.$$

- To find the slope of $\overline{KL}$, let $(x_1,y_1) = K(13,4)$ and $(x_2,y_2) = L(6,8)$:

  Slope of $\overline{KL} = \dfrac{8-4}{6-13} = \dfrac{4}{-7} = -\dfrac{4}{7}$ .

- To find the slope of $\overline{LM}$, let $(x_1,y_1) = L(6,8)$ and $(x_2,y_2) = M(-2,4)$:

  Slope of $\overline{LM} = \dfrac{4-8}{-2-6} = \dfrac{-4}{-8} = \dfrac{1}{2}$ .

- To find the slope of $\overline{MJ}$, let $(x_1,y_1) = M(-2,4)$ and $(x_2,y_2) = J(1,-2)$:

  Slope of $\overline{MJ} = \dfrac{-2-4}{1-(-2)} = \dfrac{-6}{3} = -2$.

Sides $\overline{JK}$ and $\overline{LM}$ have equal slopes and therefore are parallel. Sides $\overline{KL}$ and $\overline{MJ}$ do not have equal slopes and therefore are *not* parallel.

A quadrilateral with exactly one pair of parallel sides is a trapezoid. An isosceles trapezoid is a trapezoid whose nonparallel sides are congruent.

The length of a segment whose endpoints are $(x_1,y_1)$ and $(x_2,y_2)$ can be determined using the distance formula:

$$d = \sqrt{\left(x_2 - x_1\right)^2 + \left(y_2 - y_1\right)^2}\,.$$

- To find the length of $\overline{JK}$, let $(x_1,y_1) = J(1,-2)$ and $(x_2,y_2) = K(13,4)$:

  $$JK = \sqrt{(13-1)^2 + (4-(-2))^2} = \sqrt{12^2 + 6^2} = \sqrt{180}\,.$$

- To find the length of $\overline{LM}$, let $(x_1,y_1) = L(6,8)$ and $(x_2,y_2) = M(-2,4)$:

  $$LM = \sqrt{(-2-6)^2 + (4-8)^2} = \sqrt{(-8)^2 + (-4)^2} = \sqrt{80}\,.$$

Sides $\overline{JK}$ and $\overline{LM}$ have different lengths and therefore are *not* congruent.

$JKLM$ is a trapezoid but *not* an isosceles trapezoid because $\overline{JK} \parallel \overline{LM}$, $\overline{MJ} \nparallel \overline{KL}$, and $\overline{MJ} \neq \overline{KL}$.

| Topic | Question Numbers | Number of Points | Your Points |
|---|---|---|---|
| 1. Properties of Real Numbers | — | — | |
| 2. Sequences | — | — | |
| 3. Complex Numbers | 7 | 2 | |
| 4. Inequalities, Absolute Value | 27 | 4 | |
| 5. Algebraic Expressions, Fractions | 25 | 2 | |
| 6. Exponents (zero, negative, rational, scientific notation) | 15, 23 | 2 + 2 = 4 | |
| 7. Radical Expressions | 20, 22 | 2 + 2 = 4 | |
| 8. Quadratic Equations (factors, formula, discriminant) | 11, 24, 31 | 2 + 2 + 4 = 8 | |
| 9. Systems of Equations | — | — | |
| 10. Functions (graphs, domain, range, roots) | 3 | 2 | |
| 11. Inverse Functions, Composition | 16 | 2 | |
| 12. Linear Functions | — | — | |
| 13. Parabolas (max/min, axis of symmetry) | — | — | |
| 14. Hyperbolas (including inverse variation) | 2 | 2 | |
| 15. Ellipses | — | — | |
| 16. Exponents and Logarithms | 28 | 4 | |
| 17. Trig. (circular functions, unit circle, radians) | 10, 17, 19, 33 | 2 + 2 + 2 + 6 = 12 | |
| 18. Trig. Equations and Identities | 9, 32 | 2 + 4 = 6 | |
| 19. Solving Triangles (sin/cos laws, Pythag. thm., ambig. case) | 14, 21 | 2 + 2 = 4 | |
| 20. Coordinate Geom. (slope, distance, midpoint, circle) | 4 | 2 | |
| 21. Transformations, Symmetry | 6, 13 | 2 + 2 = 4 | |
| 22. Circle Geometry | 8, 26 | 2 + 2 = 4 | |
| 23. Congruence, Similarity, Proportions | 1 | 2 | |
| 24. Probability (including Bernoulli events) | 12, 30 | 2 + 4 = 6 | |
| 25. Normal Curve | 5 | 2 | |
| 26. Statistics (center, spread, summation notation) | 18 | 2 | |
| 27. Correlation, Modeling, Prediction | 29 | 4 | |
| 28. Algebraic Proofs, Word Problems | — | — | |
| 29. Geometric Proofs | — | — | |
| 30. Coordinate Geometry Proofs | 34 | 6 | |

**Total Raw Score =** _____

## HOW TO CONVERT YOUR RAW SCORE
## TO YOUR MATH B REGENTS EXAMINATION SCORE

Below is the conversion chart that must be used to determine your final score on the August 2004 Regents Examination in Math B. To find your final exam score, locate in the column labeled "Raw Score" the total number of points you scored. Then locate in the adjacent column to the right the scaled score that corresponds to your raw score. The scaled score is your final Math B Regents Examination score.

### Regents Examination in Math B—August 2004
### Chart for Converting Total Test Raw Scores to
### Final Examination Scores (Scaled Scores)

| Raw Score | Scaled Score | Raw Score | Scaled Score | Raw Score | Scaled Score |
|-----------|--------------|-----------|--------------|-----------|--------------|
| 88 | 100 | 58 | 79 | 28 | 52 |
| 87 | 99 | 57 | 78 | 27 | 51 |
| 86 | 99 | 56 | 78 | 26 | 50 |
| 85 | 98 | 55 | 77 | 25 | 48 |
| 84 | 97 | 54 | 76 | 24 | 47 |
| 83 | 97 | 53 | 75 | 23 | 46 |
| 82 | 96 | 52 | 75 | 22 | 44 |
| 81 | 95 | 51 | 74 | 21 | 43 |
| 80 | 95 | 50 | 73 | 20 | 42 |
| 79 | 94 | 49 | 72 | 19 | 40 |
| 78 | 93 | 48 | 71 | 18 | 39 |
| 77 | 92 | 47 | 71 | 17 | 37 |
| 76 | 92 | 46 | 70 | 16 | 35 |
| 75 | 91 | 45 | 69 | 15 | 34 |
| 74 | 90 | 44 | 68 | 14 | 32 |
| 73 | 90 | 43 | 67 | 13 | 30 |
| 72 | 89 | 42 | 66 | 12 | 28 |
| 71 | 88 | 41 | 65 | 11 | 27 |
| 70 | 88 | 40 | 64 | 10 | 25 |
| 69 | 87 | 39 | 63 | 9 | 23 |
| 68 | 86 | 38 | 63 | 8 | 21 |
| 67 | 86 | 37 | 62 | 7 | 18 |
| 66 | 85 | 36 | 61 | 6 | 16 |
| 65 | 84 | 35 | 60 | 5 | 14 |
| 64 | 84 | 34 | 59 | 4 | 12 |
| 63 | 83 | 33 | 58 | 3 | 9 |
| 62 | 82 | 32 | 57 | 2 | 6 |
| 61 | 81 | 31 | 55 | 1 | 3 |
| 60 | 81 | 30 | 54 | 0 | 0 |
| 59 | 80 | 29 | 53 | | |

# NOTES

**NOTES**

# NOTES

# NOTES